THE GOD FREQUENCY

Somewhere, something incredible
is waiting to be known.

by

Douglas Hemme

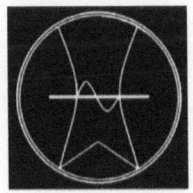

IBSN 979-8-9914671-1-7

First Edition 2024

Chapter 1

The Homecoming

The haze on the westward horizon gave the sun a horizontal slash, as if it were a ringed planet reflecting light off the ocean. A cool, salty breeze blew over the crashing waves and the squawking seagulls.

The scene reminded Antwan of a time not long ago when his biggest daily decision after surfing the waves was whether or not to ditch the afterparty and surf the radio waves on his amateur set. Over the past couple of weeks, Antwan spent his evenings watching the sun set over the horizon of the Pacific Ocean and contemplating his life's purpose. Most of his childhood friends had found jobs in retail or as tour guides or on occasion as a military enlistee.

His younger brother, Jarad, had been working at a local machine shop since graduating high school a couple of years ago, and he still lived with their parents. Antwan told himself *not everyone has the same aspiration or ambition,* but he acknowledged that at least

1

Jarad is making a life for himself and developing a skillset.

For Antwan, high school memories were already like the haze on the horizon—partly caused by the rapid tempo of his present life, obscured by the occasional party night, but mainly due to his focus on academics over the past four years. Antwan had just graduated summa cum laude from the University of Southern California (USC). He was now a degreed electrical engineer.

Antwan enjoyed coming home to Oceanside, a reprieve from the bustling atmosphere of Los Angeles and the disciplined lifestyle he voluntarily imposed on himself at USC. There were downsides to the homecoming, however. Nostalgia aside, his friends and family wanted him back home for good. Antwan's father, Byron Richard, was a retired United States Marine Corps firefighter and never wanted a life of travel for his sons.

"Why don't you get a job at San Onofre?" he often asked during trips home from college. A soft-spoken but formidable figure, his father was proud of his son's academic achievements, but he was equally eager for him to translate those into monetary success. What better place than the local nuclear power station? There was certainly a sense that working there would be a win-win situation. Antwan had fully enjoyed the educational benefits transferred to him through his father's post 9/11 GI Bill, so it would be reasonable and appropriate to consider the option, but that was not the future he saw for himself.

Antwan's mother, Maria, didn't want him to settle.

"Find your true passion and pursue it," she said. "Money can be made later. Time cannot."

A real estate agent married to her high school love, Maria had given up her pursuit of music when Byron asked her to marry him and become a military wife. She dropped out of Louisiana State University (LSU) and followed him to North Carolina when he began his military career. Byron occasionally teased her, saying he "saved her from those god-awful colors," on the side. The two spent many nights in the garage working on repairs or projects that sparked their interest. It was a bonding experience for the brothers that evolved into a means of stress relief from the other aspects of their lives.

He only had the last two months of summer to spend in Oceanside. Antwan already knew where his path would lead, even before he graduated. USC would be his home for at least another year as he had been accepted into the Ming Hsieh Department of Electrical and Computer Engineering Master's Degree program to pursue his passion and degree concentration: analog, mixed-signal, and radio-frequency integrated circuits.

Lost in his thoughts, Antwan barely noticed the vibration in the pocket of his board shorts, subconsciously dismissing it as a phantom buzz—those involuntary vibrations mimicking the familiar sensation of a cell phone. This time, it was the real deal. *Probably spam,* he thought, but he pulled the phone out of his pocket to check anyway.

It was Pedro. Antwan answered the phone call.

"Wassup, Bruh? Do you want to come over and hang out tonight?"

"Nah, I'm about to head home and have dinner with the family. My brother is working the next two nights, and I need his help with something."

"No worries. Maybe next week then. Later," said Pedro as the call dropped.

Antwan had known Pedro since he moved to Oceanside in middle school. Antwan's father, Byron, had just been stationed at Camp Pendleton and it was to be his last duty station before he retired. When the empty moving truck pulled out of the driveway of their new home, Byron quipped, "The next time I move, they'll be taking me in a pine box."

The family was all the better for it. Relentless relocations every three to four years take a toll on the family, and one of the few things they could all agree on was the desire to remain in one place. Antwan could not imagine a better place to settle than Southern California.

At every new location, Antwan would struggle to fit in. He found more comfort in schoolwork and personal hobbies than participating in the social norms for the locale. *Why bother when we'll be moving again in a couple of years,* he thought. Those were his strengths anyway. Unlike his father and brother, he wasn't built for contact sports. Antwan was average height, slender build, and would fit in better at a rave, nightclub, or a beach party than he would at a football or basketball game, especially with the dreads he kept in his hair the past couple of years.

In a fortunate set of circumstances, when he moved to Oceanside, Antwan and Pedro were placed in the same class. They quickly became inseparable. Pedro wasn't a stalwart academic like Antwan, but he was socially fluent, and he brought Antwan into his social circle seamlessly. It was the first time Antwan felt like he belonged.

The two matured together through high school, learning to skateboard and surf, living the southern Californian slacker lifestyle—as much as Antwan could—being raised in a Black Creole, military household. But he never slacked on learning. Theirs was a symbiotic relationship, Antwan and Pedro. Antwan helped Pedro pass his classes and they walked across the graduation stage together.

College was never in the cards for Pedro. Fiercely loyal to both friends and family, he had to get a real job to support his family as soon as high school ended. He wasn't resigned to working in landscaping or construction. He worked hard through summers and high school, saving his money, then coordinated with like-minded friends to open a surf shop on the beach. This took up much of his time lately, especially during the summer months.

Unbeknownst to Antwan, this summer would be the most memorable of his life. He would make the discovery of a generation, a frequency anomaly, calling into question the current understanding physics and of the universe.

Chapter 2

The Transceiver

Antwan had always been fascinated by the unseen, from the powerful winds that spun the massive blades of the wind turbines in the mountains to the east, the invisible flow of electricity through household products—a blind motive force harnessed by humankind—to the seemingly omnipresent radio waves that carried the sound to the handheld his father carried when he was on active duty.

Antwan was a high school senior when he discovered his true passion for radio waves. He was introduced to amateur "ham" radio through the Amateur Radio on the International Space Station (ARISS) at an event at the Fleet Science Center in San Diego. To converse with astronauts in space in real time was awe-inspiring, but that was only the tip of the iceberg for Antwan.

A new world had opened up to him, one that consisted of high-frequency (HF) radio wave propagation through the ionosphere, actually dependent upon real-time solar activity, or VHF/UHF repeater access from the California coastline into the Palomar Mountains and beyond.

Amateur radio operators are the true pioneers of technological exploration in radio communications. They are a diverse group of people with varied backgrounds and equally varied interests in the amateur radio frequencies from an automatic packet reporting system (APRS), the locational tracking of broadcast signals using a global positioning system (GPS), to a weak signal propagation reporter (WSPR), which is a way to track weak radio signals to see how far they travel. There are numerous ways to explore amateur radio, or one could always develop their own way to use it as well. Radio wave propagation was the most fascinating to Antwan, and it was the inspiration that led him to electrical engineering in the first place.

After gaining his extra amateur radio license the summer before college, the highest-level amateur radio license, Antwan began collecting receivers and transceiver radios that combined a transmitter and a receiver. He also acquired HF, VHF, UHF, radios and scanners. It didn't matter what modes, frequency band, dated technology, or antenna because all of these devices had their uses and features. Some more accurate than others. Some more precise than others. The radios, parts, and accessories continuously piled up in the garage where he repaired appliances. Even used, however, the most precise and most powerful models were still out of his price range.

That is when he decided he would build his own, and it would be the most precise and most sensitive transceiver in the world. Antwan didn't kid himself about this task. He knew that radio companies, signal processing companies, experimental laboratories, universities could all do this better, but they won't. Sure, companies could do it, but only if there was a market for it. Laboratories and universities could do it, too, but only if there was a research need for it, especially since they have to justify the funding.

That is where he had the advantage. He was the only one attempting an insane level of precision, sensitivity, and accuracy in measuring and transmitting radio frequencies. There was no need for this level of precision and no money in it. This was now a personal challenge for Antwan, without any challengers, save science and technology.

It was obvious at the beginning of this project that to garner a level of precision that was so impractical to control mechanically and at a level way beyond the levels in current market radios, it would have to be computer controlled. It would have to be a software defined radio (SDR). The beauty of amateur radio in the United States is that much of the software is open sourced by other amateur radio enthusiasts, and there is an online community of knowledgeable operators to assist.

The software component took a bit of sorcery. Using an off-the-shelf, Linux-based SDR program and the Internet of Things (IoT) as components, the detection algorithms had to be rewritten to account for such small-scale precision. Layering detectors scaled and filtered in an overlapping orientation much like the neutron detectors in a nuclear power plant used to monitor the level of power in the reactor core was the best way to get the desired precision and sensitivity that were better than the best in existence.

Unlike nuclear plant neutron detectors, a simple linear scale would suffice. No convoluted logarithmic scales were required, so perhaps something like a resistance decade box was a better option. Either way, that meant even more overlapping detectors. Once this receiver circuitry was developed, it was fed into the control circuitry for the transmitter with a continuous feedback loop. This was the only way the transmitted signal could be just as precise and sensitive as the receiver end.

Antwan meticulously disassembled and reconfigured basic radio components to build a robust receiver and transmitter with redundant circuits, and then he interfaced the flagship component of his build—the frequency synthesizer—with both the receiver and the transmitter. Both devices were added to the antenna output via a duplexer.

Soon after Antwan arrived home from graduation, his rig was finally complete. His unassuming conglomeration of radio parts and circuitry mounted in a worn-out server rack salvaged from a project upgrade at a local data center looked more like discarded tech waste dropped off at the entry of a Best Buy than the world's most precise transceiver.

While Antwan's radio certainly didn't rival the portability features of the top-of-the-line production model amateur radios put out by less than a handful of amateur radio manufacturers across the globe, his purpose for this one was different.

Antwan certainly wasn't planning to climb a mountain with this almost refrigerator-sized server rack filled with equipment and then run back down to bring up a generator for a field day with the other amateur radio operators. No, this was an experimental rig made of an amalgamation of radio parts to test the limits of a 22-year old's abilities before getting a formal education in design done in the institutionalized way.

Regardless of reasons for the thing's build, it was time to see if it worked. A series of 12-volt power supplies mounted side by side and stacked in the server rack gave the appearance of a makeshift bomb or homemade solar project. Wires were twisted onto terminals, tightened onto them with plastic nuts, and woven throughout the contraption like veins providing much needed oxygen and nutrients to sustain life. A pair of battery backup surge protectors acted as a foundation for the miniature shell of a skyscraper reminiscent of buildings under construction in the early 20th century.

"This is it," Antwan declared as he plugged in one thick three-pronged plug, then another. The LEDs lit up on the front displays. He pushed the two power buttons on the front.

The rhythmic yet diverse low hum of plastic-blade fans filled the tower with life. *It sounds like a beehive or swarm of drones with low batteries,* thought

Antwan, letting out an inaudible chuckle, amused by his own comparison. A touch control screen mounted on the front of the cabinet lit up brightly to prominently display the radio control software. No boot-up latency like other systems plagued the run as parasitic programs relegated to the rules and structure of the overarching operating software.

This system wouldn't be crippled by random software updates or fail to boot due to a corrupted network patch. It could not be hacked. The radio was not connected to the worldwide web. Its only interface with the world was a microphone input, power cords connected to the grid, a USB port for file transfer and firmware updates, and an antenna connection currently attached to a VHF/UHF antenna. Its only human interfaces were the keyboard, the USB port, a touch screen, and a mouse for the SDR control module.

Using the keyboard, Antwan selected FM mode and dialed in the receiver frequency to the local National Weather Service broadcast at 162.425 MHz. A faint static could be heard, barely discernible over the humming of the fans, emanating from a second-hand Boston Acoustics computer speaker, then crackling as the radio was tuned, and then disappearing just as quickly as if interrupted by the monotone narrator imbued with self-import who insists on blessing anyone who will listen with the current humidity and probably inaccurate estimate of tomorrow's high temperatures.

It is working. I've done it, thought Antwan.

But until the transmit and receipt for the frequency precision, accuracy, and sensitivity are proven, it would be the most overkill and impractical weather radio ever built.

Diving into the software, it was clear the receive frequency of the NWS station was not going to be an adequate control for measuring precision. Not just that, but Antwan knew that temperature and humidity would also be problems. Antwan begrudgingly knew the solution: The contraption would have to be placed inside a chest freezer and then he would have to build another one just like it.

Chapter 3

The Transceiver Part Deux

Just two weeks in as a degreed engineer, and I have already duplicated the last four years of radio building, Antwan thought smugly. Never mind the years of research, self-study, a shameless collection of amateur radios and electrical components collected over the years, and an already-functioning prototype. Regardless, neither of the two transceivers was a seamless build. Most parts were off the shelf, plucked from one of the many radios shelved at his make-shift appliance shop in his parents' garage. Those had just required a little cleaning to knock the dust off.

The scaled frequency detection was the most challenging piece. Occasionally, blown-out solder joints would set back the project a day, as a similar component would have to be acquired and specification differences were accounted for or compensated by additional hardware or software modification. The control operating system was initially the same as the first transceiver, but changes had to be made due to similar but different components in the second build.

A second server rack was easy enough to find at the same location, and both were the perfect size to fit into chest freezers. When mounted properly, the components should not have a problem working in a vertical orientation either. There were some components for making a second transceiver that just couldn't be found again on such short notice. Power supplies were also in short supply on the second-hand market and had to come out of Antwan's post-graduate college fund, hard earned through appliance and electronic repairs. Usually strictly disciplined with saving, this had to be an exception. The chest freezers, as expected, were easily found in the normal course of business, and soon two were ready for installation.

Antwan arrived home from his nearly nightly sunset self-contemplation. His family routines were also playing out as they did most days. Jarad had just recently awoken from a nap, a ritual shifting of circadian rhythm the day before working shifts through the night. Maria had completed showing houses this afternoon to a midwestern couple relocating to the area and they were close to a contract. Always eager to provoke conversation at the table, the starters were more enthralling than ones that could have come straight out of *Real Estate Digest*.

Byron had a hard time adjusting to civilian life after 20 years as a Marine. He still deemed the trajectory of his career to be a success, and the culmination of his final tour was treated as though it was determined by his own accord. He might have had a great influence on the decision, as he was a persuasive figure and highly regarded in his battalion. The transition to civilian work culture after decades of regimented military life did not sit well with Byron and kept a cushy corporate leadership position well out of his reach.

More than one job interview was ended abruptly, usually by the interviewee. Retirement meant he had the luxury of finding the right job, and the ability to say no, which Byron took to heart. There was satisfaction in walking away from a job that would not be the right fit, especially after years void of choice. It did not take long, though, before public service came calling again, landing him right back into management of firefighters at Camp Pendleton, this time as a civil servant.

With their busy schedules, dinner tonight was Pedro's Tacos, the self-acclaimed "World's Best Tacos Since 1986." The cartoon man holding a surfboard on the logo was an all-too-obvious comparison to Antwan's friend, Pedro Cruz. As kids, they would all tease Pedro every time someone saw it. It still amuses Antwan, thinking it would be an even better logo for Pedro's surf shop than a taco shop.

As a soft-spoken man, Byron would have been content with silence at the dinner table, despite that proving unlikely with Maria and Jarad in attendance. Antwan, on any other day, would also have let the two of them steer the conversation, but today he needed to enlist his brother in a task for which he was best suited. After waiting for a lull in the conversation between the two of them, Antwan made his solicitation.

"Jarad, I could really use your help finishing up a project. Are you free this evening?"

"Sure, but do you have anything that pays?" asked Jarad.

"Nothing we can make money on, but I can cover your time tonight," said Antwan.

"Alright, we'll make a night of it," said Jarad.

Jarad wasn't disciplined with his money, but once he realized that he wouldn't be moving out of his parent's house anytime soon, he began to pay them rent and contribute groceries to the household. He had a decent full-time job, but he knew it wasn't enough to survive on his own in the Southern California economy or its inflated real estate market.

Antwan had everything ready to go: chest freezers, power cords, surge protectors, battery backups, power supplies, monitor mounts, speakers, server racks with cards and components packed in. He just needed someone with machine skills to put it all together, and his brother fit the bill perfectly.

Tonight, they would complete not just one, but two, transceivers, a great crescendo to the years of work Antwan had put into this endeavor. Paying his brother for his expertise would be well worth the money just for the gas savings from not having to transport all of these components to a machine shop somewhere else. But it was also the main activity they shared. Antwan had already acquired a small utility trailer with antenna mounts to transport the radios.

Jarad made quick work of the job. The server racks fit perfectly in the interior of the chest freezers, a screen mount was placed on the outside of the metal housing, and the wheels of the chest freezers were upgraded to pneumatic knobby tires for shock absorption and ease of mobility. Once everything was wired up and mounted properly, the power cords were routed through the internals of the cabinet, ending with two simple three-prong plugs coming out the backside of the freezer.

Modifications were complete. Antwan rolled one of the completed units to the wall of his workshop. He plugged in a thick three-pronged plug, then another, and then another. The hum of the refrigerant cycle of the chest freezer added a new, distinct sound to the rig, one that will inevitably give it a new characteristic sound, as the tower fans were housed inside. He lifted the lid, then pushed the power buttons on the front of the first then second battery power supplies.

Again, the hum of computer fans filled the tower with life. *Too bad this beehive sound will be muffled inside this cabinet,* thought Antwan, a bit disappointed.

The touch control screen mounted on an articulating arm rising above the left side of appliance lit up brightly, prominently displaying the radio control software. An antenna mount ascended above the right side of the cream-colored lid of the device, functionally no more than an aluminum pipe holding a vertical VHF/UHF antenna with three radials extending downward, spread 120 degrees apart, just high enough not to interfere with the freezer's door.

Antwan selected FM mode and tuned the receiver to the local NWS broadcast, 162.425 MHz. This time, a clear, crisp voice broadcast above the muffled drone of the freezer's compressor. The new, shallow speakers mounted flush within body of the freezer cabinet were a stark contrast to the second-hand computer speakers of the initial build.

The aesthetics of the rig were a major downgrade. The steam punk-esque open cabinet industrial look gave way to a plain rectangular appliance with a screen, cords, and antenna, giving the rig the look of a boom-box ice cooler or a "Star Wars" droid design that was never completed. *Functionality is what matters. Aesthetics are for the marketing engineers to figure out,* thought Antwan.

So far, this has superior sound quality compared to before, thought Antwan as he dove into the software using the display monitor to check the accuracy of the receive signal. It was immediately evident that the changes had improved the reception frequency's precision. The scaled and layered filters were working as intended inside the chilled compartment. The reception quality was noticeably better from a more elevated position.

The transceivers were complete and confirmed to operate. Antwan hooked up his phone to an FM broadcast adapter and tuned both radios in to the adapter frequency. A few phone taps later and the catchy riffs of R.E.M. begin to flow out synchronously from the speakers of both freezers. So far, these were the most overkilled and impractical FM radios ever built.

Jarad and Antwan sat back in the worn-out recliners of their workshop, reveling in the shared accomplishment. Dawn was beginning to color the sky.

"I never understood frequency, uh huh...
"You wore our expectations like an armored suit, uh huh..."

Chapter 4

The Thalassophile

As a little girl, Lauren McCartney loved the ocean. So much so that she opted to go with her father, Francis, on boat rides rather than go to the nearest park where there were more things to do. There were many nights where they sat on their back patio next to the pool and watched the waves roll in across the horizon, slowly making their way to the shoreline, where at the last minute, a white cap formed on the crest of the wave, then curled forward as it made its way to the shallow beachfront.

As she grew older, she gained a new respect for the water when she began to scuba dive with her parents, and then again as an early teen when she began to surf. The ocean was her life. She was a thalassophile.

In the summer, her dad occasionally took trips for work as a marine biologist. She was fortunate to travel the world with him, experiencing the phosphorescing glow of algae in the ocean, diving through cathedrals of kelp, and witnessing columns of fish swimming past with no sense of danger by her presence.

Her mother, Cynthia, had a very successful career as an anesthesiologist that kept her busy most of the time and was not conducive to childcare. Her father, on the other hand, had the perfect job for it. It was no surprise that she chose to enter the field of oceanography.

On one such trip as a teenager with her dad, Lauren developed an interest in learning the operations of the ship they were on. She was allowed on the bridge to observe the operations, and the VHF marine radio drew most of her attention. This was their lifeline to the shore and any other vessels that were near enough for them to reach. This was her introduction to radios that evolved into an interest in amateur radio on the VHF and UHF frequency bands. She created her own radio station in the study room of her parents' house in La Jolla, and she listened to any traffic that floated across the ocean to her house on the shoreline.

In her teen years, her dad could no longer keep up with the full-time pace of field work and landed a job in academia that had a great combination of teaching and discovery research. Her mother, tired of running her own business, sold it to her partners. When the opportunity came, she took a job at the same university as her husband. The University of California San Diego School of Medicine was conveniently located near their home in La Jolla, California. This was the best life change for Francis and Cynthia and for Lauren and her parents.

This occurred at the perfect time, too, since Lauren would soon graduate from high school and go to college. She had many options to choose from, and she chose to stay in La Jolla with her parents and attend their university. Her major was a harder decision than choosing a location, and it wasn't difficult. The University of California San Diego had its own college, Scripps Institution of Oceanography, dedicated to her greatest interest: working to understand and protect the planet and find solutions to our most pressing environmental challenges.

Chapter 5

The First Contact

After Antwan obtained his amateur radio operator license, he was excited to get active on the air waves. Soon, he found a handheld radio, much like the one his dad used to carry but assigned to the amateur band frequencies with significant upgrades: color digital screen, more modes than he knew existed, and APRS tracking. Cruising the two-meter VHF band, he began to find a number of repeater channels in the area supported by local amateur radio clubs. In addition to the periodic HF events sponsored by the Amateur Relay Radio League, local groups also held weekly call-in nets, where ham radio operators gather, take a roll call, and hold discussions over the air waves.

After tuning in a few sessions, Antwan quickly realized this wasn't the place that he was looking for. He just didn't have a connection to the topics of predominately retired men, nor were the times convenient to his school and work schedules.

He then wandered the frequency band looking for a simplex radio contact, one that doesn't shift transmit and receive frequencies to go through a repeater.

"Is the frequency in use? This is N7APS."

"CQ, CQ, calling CQ. This is N7APS, November-Seven-Alpha-Papa-Sierra."

He knew this action to be futile. A handheld radio on two-meter band with a maximum of five watts output would not have a great range without using a repeater. Still, he kept at it just in case.

"Is the frequency in use? This is N7APS."

"CQ, CQ, calling CQ. This is N7APS, November-Seven-Alpha-Papa-Sierra."

Sometimes, he would adjust the squelch and just listen to the static. The noise came through the receiver and all the circuitry within the radio. Then the occasional conversation came through, as if the callers' frequencies shifted in the wind in tune and out of tune from his vantage point. Still, he would occasionally send out his own call, just in case:

"Is the frequency in use? This is N7APS."

"CQ, CQ, calling CQ. This is N7APS, November-Seven-Alpha-Papa-Sierra."
Antwan broadcast one more time over the air on an arbitrary frequency mid-range of the two-meter band.

"N7APS, N7APS, this is KJ7EBU, Kilo-Juliet-Seven-Echo-Bravo-Uniform," a faint soothing voice swept into the room from the radio speaker like the cool ocean breeze wafting in through the curtains of Antwan's open window.

At last, a first contact.

Dashing down the Pacific coastline, from Oceanside to La Jolla, skimming across the surf, dodging the occasional seagull in line of sight, a sine wave at a VHF two-meter band radio frequency flew through the air at the speed of light and with it a carrier signal modulating to the voice of its sender. Lauren McCartney, a second-year oceanography student at University of San Diego, was on the opposite end of the experience.

Though Antwan was two years younger than Lauren, her and his interests were not limited to amateur radio: academics, technology, renewables, surfing, sailing, the ocean, the environment.

Regardless of their rigorous schedules, the two found times for their own meet-ups on the air waves. This was the beginning of a continuing long-term friendship.

Chapter 6

The Discovery

Now that the transceivers were complete, Antwan needed another ham radio operator with a fixed radio station to help him calibrate the two devices. There was only one person he trusted in sharing this task: Lauren, KJ7EBU. One problem was they hadn't actually met in person.

Antwan scheduled a meet-up with Lauren on their usual frequency, which was their most comfortable mode of communication. A text was going to be too long to explain, and they just didn't do voice calls.

He needed to ask her to meet in person and if she was willing to help him with his project. No, this was too important for an impersonal text. For four years of friendship, they commiserated on graduations, discussed stresses of exams and finals, consoled each other after losses, and shared life goals, but they had never suggested meeting in person. Perhaps they were the blind leading the blind, or they took comfort in their separate lives sharing parallel academic successes, but meeting never came up until today.

Lauren had just graduated with a master's degree in geophysics in route to her PhD. Most of her time was devoted to her academic goals. She came from a well-off family in La Jolla, as far as Antwan could tell. There were indications of such in their conversations, not to mention the fact that she lived in La Jolla. Both of her parents were doctors, her father a marine biologist and her mother an anesthesiologist. She was an only child, but had a nanny briefly while growing up. *Along with the age gap, yet another reason she is out of my league,* thought Antwan.

Still, she was the best option to prove his transceivers' precision. *This could be a significant technological contribution to the amateur radio community,* Antwan mused proudly. And he wanted not just to boast, but to include his friend Lauren in this moment.

At their scheduled meet time, 2:00 p.m. on Tuesday, Lauren and Antwan established communication on their usual frequency in the two-meter band.

"KJ7EBU, N7APS calling," Antwan continued: "KJ7EBU, Kilo-Juliet-Seven-Echo-Bravo-Uniform, this is N7APS, November-Seven-Alpha-Papa-Sierra, calling:"

"N7APS, this is KJ7EBU. How are you today?" Lauren responded.

"I have a big ask," Antwan declared, then attempted to summarize his project and what specifically he needed from Lauren.

Much to Antwan's surprise, Lauren was excited to help and equally enthusiastic about meeting him in person. All this time, he thought their friendship was only able to flourish as it did within the confines and safety of radio telecommunication. Antwan went a step further and invited Lauren out to lunch in appreciation of her preemptive assistance.

Now that she agreed to assist and plans were made to meet at 11:00 a.m. on Wednesday near Lauren's house, Antwan began to feel the twitch of nervousness as the reality of the situation sank in.

Antwan would load one transceiver and deliver it to Lauren. They would then communicate over the phone to coordinate dialing in the frequency to each transceiver and use the continuous wave (CW) function to transmit a stream of radio frequency energy to test both radios' precision, sensitivity, and accuracy. Amateur radio operators typically use CW mode to transmit Morse code over the radio waves. It seemed to be the best option for this test.

The next day, Antwan had one transceiver plugged in and ready to go; it just needed to be powered up. Rather than plug and unplug all three cords of the contraptions, he installed a small breaker to a receptacle in his workshop to control the power to the entire radio. This was just a temporary setup, though. Leaving power off to the rig would only drain the batteries and the radio wouldn't turn off without a fault or when the batteries go dead.

The second transceiver was wheeled onto a utility trailer and strapped down. The antenna and control screen were removed for travel, and the articulating mount was retracted. Antwan backed up his Hyundai Santa Cruz XRT and coupled the trailer.

He knew it wasn't, but Antwan couldn't help but ponder, *why does this feel like a Tinder date?*

It only took Antwan about half an hour cruising down "the 5" to arrive in La Jolla from Oceanside, even with the trailer. The hard part was finding parking. A couple of circles around the block and voila: a long enough spot on the curb to park his truck and trailer. *Good thing I planned ahead,* thought Antwan, leaving early to account for traffic or complications.

With no additional time to dwell, Antwan ran across Prospect Street to their rendezvous location: The Spot. Walking in the restaurant as the clock struck 11, he suspected Lauren would already be there waiting. Scanning the tables, it wasn't hard to identify the single occupant table, one blonde woman sitting patiently, head in the menu with a lackadaisical look as if the menu was something she had seen a thousand times. Considering the proximity to her parents' house, she probably had.

Her sun-kissed face contrasted with her light-colored hair, shielding her face from the public as best it could. She was wearing casual clothes: jean shorts and a white blouse with sandals. A California girl, for sure. None of this mattered, but Antwan was making a mental note of everything in an effort to reconcile what he knew of this person with her appearance. Everything seemed in alignment. This was the Lauren he had been communicating with for years.

As he approached the table, a moment of paranoia hit Antwan, as if for the first time he had a deep-seated value for his own appearance. He hoped she likes slightly younger men with the looks of Juice Wrld without the money and fame, but with the intelligence of, well, Juice Wrld without the artistic flair. *A West Coast version,* he thought with a smirk.

Lauren looked up, saw him, and instantly smiled—a great indication that she didn't think she was getting catfished. Why would she? Afterall, this was a platonic meet-up at his request. It wouldn't be appropriate to change the rules after the game started.

The two college post-grads ordered food and shared stories of recent events in their lives like old friends catching up after a long absence. Lauren ordered a salad for lunch, while Antwan ordered the spicy jambalaya pasta, justifying the purchase by not having eaten all day. It had become an ongoing personal challenge for him to sample all things Creole to find any food that could rival his mother's. The jambalaya pasta had potential, but the spices just weren't the same as at home.

Eventually the conversation was about at the main topic. Not so coincidentally, this was a different personal challenge of Antwan's: transceivers.

Lauren had walked to The Spot restaurant from her house, so she got in the Santa Cruz and directed Antwan to her house. An inconspicuous gate with a long driveway past a street-facing house led to her residence. Her ocean-side house had unobstructed views of La Jolla beach from a cliff-side vantage point. The Coast Walk Trail crossed behind and below her backyard. It truly was a beautiful location, and ideal for radio reception along the Pacific coastline.

She clicked a button on a remote and the two gates opened. The driveway was long and narrow, but Antwan was able to back the trailer in, fearing the clearance just wouldn't be there to turn around later. They unloaded the transceiver freezer, which really needed a better name, and rolled it into Lauren's radio station setup in the garage.

Antwan's amateur radio station had grown with time, but it was surpassed by Lauren's. As the daughter of a marine biologist, she grew up familiar with VHF marine radios. Her setup was very suited to the project at hand.

They had a plan. Starting with their preferred meet-up frequency, they would dial down deeper and deeper into the frequency and compare the transmit and receive frequencies with the transceivers and additional wideband receivers in their radio stations. Alternate, verifiable indications were very important.

Antwan helped set up the transceiver and showed Lauren the build from power input and battery backup power supplies all the way through to the user input, control screen navigation, and the components of the antenna. She seemed quite impressed by the setup. It certainly wasn't something she'd seen before, which is understandable, considering she devoted her time to other endeavors instead of taking apart radios. Most of the radios in her station were either bought new or were well cared for—a big difference from the ragtag group that lined the shelves in his garage.

Antwan said his goodbyes, and they scheduled a time to start testing that afternoon. At the appointed time, they fired up the transceivers with frequencies lined up to 145.4500, being careful to avoid the frequencies of the local two-meter band repeaters on the higher end of the band. Finally, the testing had begun.

Lauren and Antwan took turns transmitting and receiving a CW signal, narrowing the transmit and receive frequency to match and recording the frequency detected by both the hand-built and commercial manufactured receivers. The transmit and receive power was also recorded.

After our face-to-face meeting, it didn't seem so awkward a concept to actually talk to Lauren on the phone, Antwan mused, *if we only had met up sooner.*

The first transmissions were a success and right on target. There was a definitive drop in power between the transmit and receive power levels, as expected from a distance of >30 miles. Dialing in even longer integer frequency values was an arduous process, reminding Antwan of the time he accidentally closed a 3-digit bike lock without knowing the combination. Having all but given up before he started going through the possibilities, he was adamant he would not go through 1,000 iterations of trial and error to get the right value. That is when Jarad stepped in and began: 0-0-0 click, 0-0-1 click, 0-0-2 click. Scoffing, Antwan turned to leave the room, then 0-0-3 click click. The lock opened! His brother had unlocked the bike lock in three simple steps. If there was a moral to this story, it would be don't give up even in the face of overwhelming odds.

Antwan and Lauren had developed a process to randomly pick numbers between 0 and 9 to serve as the next integer when descending into the abyss of the frequency metering 145.45258...579 Hz.

The actual numbers didn't matter to Antwan so long as they matched. And so far, everything was working great until everything was not: 145.45258...579525 MHz, out to 16 significant digits, then 20, accurate to even less than one attohertz, aHz, or 10 to the minus 18 Hertz, the standard unit of frequency equivalent to one cycle per second.

Then out to 24 significant digits beyond the decimal, accurate to one yoctohertz, yHz, one septillionth of a cycle per second.

This time was different.

Indicated receive power went through the roof, nearly blowing out the built-in speakers of Antwan's transceiver in Oceanside. The frequency registered as a spike on the home-built transceiver, but nothing showed on the display of the wideband receiver.

"Did you increase the transmit power?" asked Antwan, bewildered.

"No, why? Do you see the signal?" replied Lauren.

"The receive power jumped to about 50 watts, but it's only showing on the freezer transceiver. Nothing on the wideband. Perhaps a front-end component blew out in my radio or an errant signal came in from somewhere else and caused both receivers to malfunction. Why don't we stay on this frequency and I'll transmit to you?"

"Copy," she replied.

Antwan watched the freezer display as the receive signal dropped to zero, then keyed his own transmitter, sending a 50-watt continuous wave signal right back.

"This is incredible," said Lauren, seeing the same indications. "Let's shift the least significant digit up one, then down one."

"Sounds good. I'll send it," said Antwan.

A moment later, the signal strength dropped off to normal and showed again on the wideband display. It came right back up to 50 watts on the freezer and disappeared from the wideband, then came back down and displayed again. Antwan had increased the frequency by one, came back down through the original frequency, then to dropped it by one once more.

"There appears to be an anomaly occurring at that specific frequency," Lauren informed Antwan. "It just doesn't make any sense."

Antwan recorded all the settings from his device and stored the data on a jump drive for posterity. They could have stumbled upon something significant here and he did not want to lose it.

Unfortunately, life obligations got in the way of continuing experimentation, delaying their work by a few days. Antwan told his family he may have come across something, but alluded to it in vague terms, not just because he didn't know all the details, but because it would be mildly interesting to them at this point anyway.

For Lauren and Antwan, the next step was to increase distance and add line-of-sight obstacles between the transceivers. With no signal loss between the two last time, they could perhaps find a degradation in signal under more challenging conditions. Or perhaps they would not be able to find a signal at all. After all, a normal VHF signal requires a repeater to travel further distances. Without it, two radios normally cannot communicate in simplex mode at long distances. But that is precisely what they were to test here on an extremely precise frequency.

Antwan was happy to have met Lauren in real life and was enthusiastic about spending more time with her, but their current arrangement was keeping them hundreds of miles apart. *It would figure that I meet a girl and drive her away intentionally,* he lamented.

On Sunday, Antwan loaded up his home transceiver onto his trailer, along with a small Honda inverter generator. He was going to drive out through the mountains, and he and Lauren were going to test this anomaly. Perhaps there was a reasonable explanation.

Lauren's parents had a lakeside vacation home on Big Bear Lake. She offered it to Antwan to stay the night if he ended up out too late. It was either this or someone was getting on a boat, and neither of them wanted that much risk to the transceivers.

The first stop for Antwan was the base of the mountains at Chino Hills State Park. It was on the opposite side of the mountains as La Jolla, and clearly one of the least ideal locations for two-meter frequency transmission. If the signal failed to receive, they would be done with this fool's errand and he would return home. Antwan fired up the generator, making its distinct low humming sound, then turned on the output, energizing the freezer and radio components inside.

While allowing the components to cool for a moment, Antwan called Lauren and prepared for the test. When they were both ready, Lauren transmitted 40 watts on the anomaly frequency. Antwan's rig lit up, spot on at 40 watts.

"Unbelievable," he said.

Lauren, knowing what it meant, communicated changes to the power output and a shift in frequency up then down through the suspect frequency and slightly below. All indications matched like previously. For good measure, they confirmed no signal could be detected on the frequency or side frequencies from their wideband receivers. Lauren then turned off her transmit, and they repeated the tests with Antwan's transceiver, receiving all the same indications as before.

It was clear that there was no indication of signal loss from the distance or from the obstruction of the mountains. They decided that he should go straight to Big Bear Lake and try again.

Antwan made the drive in just a couple of hours, thankful they decided to perform this joint effort on a Sunday. He turned onto the unimpressively named Lake Drive, where the rows of wooden cottages and two-story homes lined the streets that smelled like pine trees. On the right side were larger homes that backed up to the waterfront, and the left side was lined with the more modest cottages, lodges, and cabins.

Lauren's family vacation home was one of the two-story homes on the waterfront. The red brick home with outward facing bay windows painted white had a downward sloping driveway big enough to back the truck and trailer into without blocking the road or colliding with the two side-by-side single-car garage doors in the front of the house. Most of the neighboring houses did not have such a long driveway or a driveway at all for that matter. Another reason to be appreciative of this day.

Lauren shared the codes to the garage and security system with Antwan, who was surprised at the level of trust she imparted on him after having met her in person just once. Antwan did not intend on staying the night in a vacation home owned by the parents of someone whom he barely knew, especially given his demographics. *Perhaps I should have borrowed one of Jarad's work shirts,* Antwan thought to himself, as he unloaded the chest freezer from the back of the trailer and wheeled it into the garage.

On initial impulse, any amateur radio operator would seek a high, unobstructed point for optimal radio wave propagation, but this was no ordinary case. Antwan had initially planned to wheel the transceiver to the back of the house where he would have an unobstructed view across the lake. He had also initially planned to mount and raise the vertical antenna. Standing in the garage, he decided not to do either one. He grabbed the power cords, extension cord, found the nearest plug, then closed the garage door. *Best not stand in plain sight for too long,* he thought.

Coordinating with Lauren, the two repeatedly tested the transmission, reception, and signal strength. Transmitting at the anomaly frequency without an antenna, the broadcast was received in La Jolla at the exact same power it was transmitted.

This time, Lauren removed the antenna from the coastal transceiver, again spawning absolutely no change to the detection or transmission signals.

This is just getting ridiculous, said Lauren, thinking to herself. *Perhaps I am being played for a fool.* She then pulled up the security cameras at the Big Bear Lake vacation home, confirming that Antwan was indeed where he said he was.

Lo and behold, staring back at her was the tailgate of a California sand-colored Santa Cruz attached to an empty trailer, just as expected. Antwan loaded the transceiver back onto the trailer and returned to Oceanside.

Today's experiment was an interesting development, further deepening the mystery of this frequency anomaly. They would have to regroup and think about what to do next.

Chapter 7

The Solicitation

Antwan invited Lauren to meet him at his humble garage workshop to brainstorm the next steps. She quickly accepted, saying she wanted to see the spot where he built both transceivers. He continued to be surprised at her interest ever since he propositioned her to help him with this project. It would be a shock to his parents as well. Antwan didn't bring women home to meet his parents and seldom brought anyone at all.

When Lauren arrived, he quickly escorted her back to the shop in a vain attempt at avoiding embarrassment, but at least a brief reprieve from the inevitable family attention bomb and to have a few minutes to converse in private.

"The next logical step to test the limits of this frequency anomaly would be to get another ham operator to confirm the signal, perhaps even one in another country," Antwan said. "But no one is going to have the capability of dialing in to this frequency," he continued.

"Maybe we should reach out to one of the universities. My parents are both faculty at UCSD. You surely have some connections at USC as well," said Lauren.

"I don't want us to get ahead of ourselves, but this could challenge our understanding of energy wave propagation at the least and maybe even quantum physics," said Antwan excitedly. "But we have to reach out to someone who is going to be able to corroborate what we have seen."

Lauren nodded her head in agreement. "So, maybe a radio observatory?" Lauren said.

Antwan responded with a head nod of his own. "We will need to reach out to someone who will take us seriously, perhaps even present our case, before gaining access to a government or university radio telescope. There is another option," said Antwan. "Amateur radio astronomers."

"In the Society of Amateur Radio Astronomers (SARA), members are mainly amateur radio operators. Perhaps we can get someone there, some distance away from us, to listen to our frequency. They'll have to have extremely precise hardware. This might be directly opposed to the wideband listening that they normally do. Maybe that's why this hasn't been discovered earlier," said Antwan.

"If we can't get anyone else, perhaps a longer road trip is in order. Regardless, we need a third set of ears."

Their private conversation came to an abrupt end as they were descended upon by a pack of attention-seeking relatives. Maria invited Lauren to stay for dinner. Byron started with the Big 4 inquisition of family, education, location, and occupation. Jarad was present just to revel in the awkwardness that the situation was sure to bring.

Lauren maneuvered the onslaught flawlessly. After dinner and a home tour, the two made their leave and got back to business. Antwan had found a contact,

Daniel Sullivan, on the radio-astronomy.org website, who just might have the hardware needed for confirmation of their anomaly.

As a retired physics professor and amateur radio astronomer, Daniel had both the expertise and time on his hands to help them. In northern Arizona, he was not as far away as they had hoped, but significantly farther away than a normal VHF frequency would travel undegraded and unassisted. Lacking any better options at this point, Lauren agreed they should reach out with what they know and await a response.

Opening his personal email, Antwan sent a message to Daniel. Rummaging through his bag, he found the thumb drive with the frequency and attached the file. Antwan then went into a high-level detail about his radio build and into minute details on the transmit and receive indications they were getting; most importantly, the distances and the terrain that existed between the two transceivers. Lauren helped Antwan with some key points, and when the two thought there was an adequate explanation for what they had witnessed, they crafted a request for assistance in further exploration of the frequency anomaly.

"I don't have the time or resources to make more of these right now," Antwan said as he gestured to the chest freezers resting along the wall of his workshop. "But we need this level of frequency precision to adequately monitor and test," he continued.

Looking at the schematics and parts scattered across his workbench, piles of printed operating manuals from various vintages of commercially made radios, and the pile of excess parts in a bin by the back entry, Lauren had a good idea of how best Antwan could use his time while they awaited a response.

"Perhaps you should formalize your design," she said. "Build schematics and CAD drawings or sketch the layout. Detail the specifications required for the precision needed to detect this frequency so someone else can replicate it. You'll need it to apply for a patent as well."

"I'll get to all of that," Antwan reluctantly affirmed. "What I would like to do is explore the transmit ability on our frequency."

"Our frequency?" Lauren replied with a smirk, her cheeks flush.

"We started with our frequency, where we've met up and talked for years now, then drilled down until we found this," Antwan said. "So, yes, I'd call this our frequency. It sounds better than the 'frequency anomaly,' That shizz is getting old."

Antwan walked Lauren up the driveway back to her car. He didn't want this to end, but was anxious to get on with the next phase of his 'project' while they waited for a response. Also, they needed a Plan B if Daniel didn't get back to him, which would most likely be repeatable design documentation for someone else to take and build their own, just like Lauren had suggested.

"I'm glad you invited me over. I am impressed by what you've done and the space you did it in," Lauren said. "And your family is wonderful."

Visibly cringing at what had transpired with his family, they both laughed.

"Thank you, Lauren, for agreeing to help and for all that you've done," Antwan replied, genuinely appreciative. "I could not have done this without you," he continued, in reference to discovering their frequency and testing his transceivers.

There was still work to be done, but they had already discovered something incredible. They hugged and he opened her car door, allowing her to drop into the narrow seat of her BMW convertible, then closed the door with just enough force to latch. *Smooth*, thought Antwan, never considering he could have sought more affection or taken things further. His focus was on the work and the discovery.

Lauren committed to meeting up in a couple of days if they did not hear from Daniel, and sooner if they did. Antwan watched as she backed out of the driveway and her taillights faded down the street. *Now, it is time to get back to work,* he thought.

Already late in the evening, Antwan didn't want to waste a moment of time. He set to work organizing his schematics, drafting new ones to match the as-built transceivers, and compiling the related portions of the various vintages of radio manuals he had collected.

Taking a break from the mundane documentation work, he set up his transceivers at opposite sides of the workshop and began exploring the capabilities of their frequency. Using the microphone input, he attached a Bluetooth adapter to one of the transceivers and tried to broadcast in FM mode. There was dead silence on the receiving end. Shifting the frequency down one digit on both, the reception came through unimpeded.

Wild, he thought. Frequency modulation won't work because the carrier signal distorts the frequency too greatly, taking it off the mark. "Of course!" Antwan yelled to the invisible audience witnessing a genius at work, "Amplitude modulation will keep the frequency locked in," he continued, as if all of the radio carcasses were listening to his revelation. Antwan changed modes on both transceivers to AM, 1 watt output, and, opened his phone and pulled up The Weeknd's album, *Dawn FM*, selected with intentionally ironic intent.

The familiar warble of electronic sound, birds chirping, came through from the opposite side of the workshop. The organ sounds of the intro emanating through space.

"This part I do alone.
I'll take my lead.
I'll take my lead on this road."

Me, too, thought Antwan. The companionship was nice, but for him this was still a comforting thought, and he was proud of what he had accomplished by himself.

Chapter 8

The Professor

Daniel Sullivan didn't get out much. One could argue, and he does, that he was always out and didn't get in much. In either case, his daily personal interactions were very limited since he moved to northern Arizona. A homestead located in a forest of Juniper trees and volcanic rock, 10 miles up a dirt road from the old Highway 66 is where the retired professor called home.

In his final days at the Polytechnic Institute of New York University, he knew how he wanted to spend the rest of his days, and it was the exact opposite of what New York City had to offer. Population growth and human development had increased beyond the pale on the East Coast, and upon retirement, there were no more ties to keep him there.

Light pollution, or lack thereof, would lead Daniel to northern Arizona. The Bortle dark-sky scale, a nine-level numeric scale that measures the night sky's brightness caused by light pollution, guided him. The lower the number, the better for observing celestial objects in the night sky.

Light pollution maps showed large pockets of Bortle Class 1 skies throughout the West. Unlike the Bortle Class 8-9 in New York City, the West was a dream for an astronomer, amateur or professional. Naturally, human development was inversely proportional to dark skies, so finding a Bortle Class 1 location was a win-win for Daniel.

He found his piece of the American dream on an off-grid homestead website. Positioned three quarters of the way up a mountain, or what at least what they call a mountain in northern Arizona, was a 40-acre plot with a decent sized rustic cabin nestled on level ground, surrounded by acres of private lands, checkered by acres of public land, and no neighbors for miles. But also no water, electricity, or internet, and spotty cell service.

Daniel had no intention of living without the comforts of modern society. Just a quarter mile away from his newly acquired property, at the mountain summit, was a microwave communications site. He reached out to the local utility, coordinated, and funded the installation of power poles tapped off of the distribution line to the communication site and on to his property.

His would be the only homestead on the grid for miles. The topography of the site also provided perfect conditions for a wind turbine most of the time. That was his backup, but there was always the less-desirable, noisy generator if needed.

For the internet, he opted for a satellite. Cellular service was not consistent enough for the internet, and it was 4G if he was lucky. A rainwater collection system was already installed, as was a septic tank, but fresh water still had to be hauled in periodically when his water supply dwindled. This was the main obstacle to any long-term self-sufficiency.

Daniel built out his personal observatory, eventually expanding into radio astronomy. However, he discovered that the nearby communications site caused major interference. Not to be deterred, he set out to find a solution.

After much research and trial and error, he completed a layered filter system to narrow the observation of radio waves, similar to Antwan's, and designed only for reception, of course. Daniel then wrote a piece on the project and shared it with the Society of Amateur Radio Astronomers. That is how Antwan discovered Daniel Sullivan.

After a late night of stargazing with a waning crescent moon, Daniel woke up late in the morning, a sliver of sunlight peeking between blackout curtains, highlighting specs of dust defying gravity in his bedroom, giving the vertical plane of illumination the illusion of a material substance. So much for blocking out light if you don't close the curtains all the way, lamented Daniel.

The dust was an unfortunate cohabitant during high desert summers, except after an occasional monsoon thunderstorm that transforms the dust into a paste that cakes onto tires in layers, filling wheel wells of vehicles, and any cracks in anything or anyone. Still, the rain is a welcome reprieve and lessens the burden of hauling water up the mountain.

Daniel started his coffee maker, pre-staged for morning convenience. He didn't mentally function as well when the sun was out. *May as well live in the desert when my best hours are during the night,* he thought as he dusted off his laptop and lifted the screen. A robot vacuum scurried across the hardwood floor of his cabin, seamlessly transitioning over a rug in the main room and around the periphery of a sleeping blue tick hound laying in the center, not in the least bit fazed by the robotic contraption.

Daniel performed his ritual skimming of the news on the internet browser's main page, reinforcing his decision to seclude himself on the mountain. Perhaps that's confirmation bias, but it's enough to keep from regretting the move. No news he can't do without.

Coffee done, he came back to his laptop to complete his ritual internet rounds, but abruptly changed his mind and closed the laptop screen. He'd check his email later. He was going to town today. *Groceries don't deliver themselves, at least not out here anyway,* Daniel thought. He got ready, grabbed his wallet and keys, then headed toward the door.

"Come on, Dusty," Daniel said. His loyal companion perked his head up, then followed Daniel out the door.

The sun chased the horizon, clouds churned into themselves as they passed across the sky, hours ticked and the sun turned the horizon's hues pink and orange, then dark blue and light blue hues as the stars begin to light the night sky.

That was the point when Daniel's RAM 2500 towing a water trailer, audibly and violently sloshing around its liquid contents, made its way up the dirt road to the automatic gate installed at his driveway. Daniel had installed and powered the gate from a solar panel and battery, but he did not shy away from running his own cables. He had installed two large dishes to create a tandem radio telescope: one on the east side of the mountain where his cabin was located, and another on the west side beyond a secondary summit under the morning shadow of the communication towers.

With chores done, dinner made, and groceries put away, Daniel still had another hour before the moon would clear the horizon. He opened the lid of his laptop to complete his ritual. Among the routine emails, bills, group newsletters, and the occasional former colleague update, there stood an outlier from the pack: Subject: Frequency Anomaly.

Daniel took a moment to read, then re-read key parts of the email. Determining that this was not in fact spam or phishing, he opened the attachment. *Woah,* he thought to himself as he saw the numbers scroll across the screen. Yoctohertz wasn't a level of frequency precision he had attempted because there was not valid reason for him to do so. He believed that he did have the hardware and software capability to tune it if he desired.

He was still skeptical, as these kids had surely made an error somewhere in their method. Perhaps they weren't that far away, or a repeater boosted their signal in simplex. That was easy enough to do. It would be impossible to calibrate a frequency detector from an identical, unproven, and uncalibrated transmitter. Regardless, he intended to tune in for himself.

Daniel poured a glass of whisky on the rocks and carried his tumbler out to his back porch and up a dimly lit pathway to the dome that houses his optical telescopes. This was his workshop. The radio telescope controller, all computerized, was located inside his observatory.

He copied the file onto a thumb drive and transferred the frequency anomaly values into his software program. He could hear the whirr of the turret on his west-facing satellite dish as the software shifted the direction and angle to point toward southern California. Daniel set up the control software to disable a dish if it could not point to target, or enable both dishes if the target was directly overhead and both dishes had a line of sight.

It shouldn't matter where its facing, from what these kids describe, he thought. But it was already part of his routine for peering deep into the night sky.

Nothing.

As expected, there was nothing on such an extremely narrow frequency band. Regardless, he left it where it was, listening, as he re-read the email, trying to come up with a reasonable explanation for what they had witnessed.

As he was lost in thought of potential sub-ionosphere wave propagation of VHF radio waves, a steady tone with an equivalent signal strength of exactly 1.0 watt appeared on the display. He was incredulous. Surely, this was a coincidence.

As quickly as it came, it went. Then it came back again, but different this time. There appeared now to be a slight deviation to the signal strength. Amplitude was changing. "AM!" Daniel exclaimed.

He navigated the software, selected AM mode, turned up the volume, and heard The Weeknd's Dawn FM album, *Gasoline*:

> *"It's five AM, I'm nihilist,*
> *I know there's nothing after this...."*
> *"Obsessing over aftermaths...*
> *Apocalypse and hopelessness"*

Chapter 9

The Collaboration

Daniel stared, wide-eyed and dumbfounded at what he was hearing. A 1.0-watt AM signal broadcast completely on a frequency narrowed down to 24 significant digits beyond the decimal. Shifted right or left by just 1×10^{-24} Hz and the signal is completely gone.

"Just incredible!" he exclaimed aloud to himself and his dog.

He then remembered that Antwan described how the signal came through even after removing the antennae. *That will require some wrench work,* he thought, *but I can keep on the frequency and switch to my east-facing radio telescope.*

A few clicks later and the same music was clearly playing from the opposite dish, arbitrarily pointed to the northeast toward the Navajo reservation. *May as well go stereo,* he thought, as he positioned both radio telescope dishes straight up, bringing in reception from both dishes. Still, crystal clear sound with equal, 1.0-watt signal strength on both.

Daniel replied all to the email Antwan sent, indicating he had heard the music playing and gave times, signal strength, and indicated it was irrelevant to the position of his radio telescopes. He suggested they do an internet call the next afternoon.

Antwan shared the log of his broadcast times from the controller of his transmit radio, which matched what Daniel had received. He and Lauren both agreed to the online meeting at 2:00 p.m. the next day.

Lauren arrived at Antwan's workshop around 1:00 p.m. As adverse as he was to be meeting new people, Antwan was excited to bring more expertise into the fold. That, and he was relieved they would no longer have the goal of seeking further distances from each other to perform further tests of his transceivers and their frequencies.

Daniel started the meeting around 2:00 p.m. from the balcony of his two-story cabin. He was proud of his property and the improvements he had made to it, with minimal intrusion into the natural landscape, except for the inevitably obtrusive radio telescope dishes and the wind turbine. After he purchased the property, he had managed to do a lot without clearing out too much more of the natural vegetation.

Lauren and Antwan logged into the meeting with Daniel using a webcam at the workbench in Antwan's workshop. They fully intended to go through the build with Daniel in detail and fully collaborate with him in exploring and sharing this discovery with the rest of academia.

Antwan was impressed that Daniel dialed into their frequency so quickly after he sent Daniel the frequency. The reason he reached out to him was his incredible work and ability to communicate the process so well in his presentation materials on the SARA website.

"Hi, Mr. Sullivan. My name is Antwan."

"And I'm Lauren."

"Pleasure to meet you both. Please call me Daniel. You have got to be proud of this discovery of yours and the transceivers you've built."

"Those were all done by Antwan," Lauren interjected, not wanting to take undue credit.

"Well, you did great work, Antwan, and thank you for reaching out to me."

"Of course, and thank you," said Antwan. "We were hoping you could help us make sense of this, because it doesn't seem to fit my understanding of radio wave propagation."

"Well, from what I've heard and seen firsthand, this may well challenge my understanding of physics, too, and I've spent my entire career in the field," said Daniel.

"You two had the right approach. Distance is what we need to prove the limits of this."

Laren asked, "If you're getting at what I think you are, are you saying that you want to measure the speed of the radio waves at the frequency of the anomaly? We could travel to the antipode, the opposite side of the world, from one transceiver with the other. That would give a distance of approximately 7,900 kilometers through the Earth, where we could expect to have a measurable 0.02 seconds going directly through the mantle and core, or would it take longer to get around anything non-permeable? What do you suggest?"

"It is a good idea, Lauren, but I would instead like to appeal for academic help in developing research into this phenomenon. At the present time, I cannot be sure anyone else has the capability of targeting this frequency. There's been no need for that level of precision in this frequency bandwidth. With your permission, I would like us to make a TED-style presentation and present it as a solicitation for research."

Daniel continued, "We will not advertise the frequency. Instead, there is a science and technology conference next week at Lowell Observatory in Flagstaff in collaboration with Northern Arizona University. I have some contacts there and could get us on the schedule. We are going to need industry support for technology development and funding for research to properly test the limits of this frequency. These projects have long lead times, but we could possibly get a transmitter on the ISS or incorporated into a satellite or probe and truly test this frequency," he continued.

Antwan and Lauren looked at each other and nodded in agreement.

"What do you need from us?" asked Lauren. "Looks like we'll get to take that road trip together, Antwan," she said, shooting him a sly grin.

"We've got a stop to make first. I would like to give a live demonstration," said Antwan.

Daniel, true to his word, procured a 20-minute presentation spot at the science and technology conference next Thursday, under a vague topic header: Radio Frequency Phenomenon. He tasked his new associates with providing a high-level overview of the transceiver build as well as demonstrating their discovery, being careful not to divulge much, if anything, regarding the frequency band or the frequency itself. It would then be Daniel's time to finish with the potential implications of their finding.

The next day, Antwan and Lauren met up with Pedro at his surf shop on the boardwalk. Antwan explained everything to him and asked if he'd help out with their demonstration. He was game. Antwan wheeled in one of his freezer transceivers and showed him all he needed to know to broadcast at the right time, and he set him up for the video chat.

Pedro found it hilarious that Antwan had made a radio out of a chest freezer with an antenna, screen, and big wheels.

"You should change your call sign to R2D2," he joked.

"You're a fool, man," Antwan replied. "But I love you anyway. Thanks for doing this."

Antwan and Lauren met Daniel at his cabin on Wednesday afternoon. They had already crafted a presentation to open the talk and were eager to see Daniel's observatory where he captured Antwan's signal. There was plenty of room in the backseat of Antwan's truck for their luggage for this brief trip after he removed the wheels of one transceiver to place it in the bed of his Santa Cruz. He was thankful he had strapped it down securely before making their way down the long dirt road off of Highway 66 to Daniel's property.

As they approached the gate to this property, Antwan wished that they had coordinated radio communications because he didn't have cellular service. That thought quickly became moot as the gate opened as they approached. They could not see the house through the trees, but noticed there was at least one security camera inconspicuously hidden in a branch.

Lauren and Antwan made their way up the dirt drive lined by juniper trees and over a culvert, winding upward to a clearing in front of a modest cabin that looked like it grew out of the earth rather than being built on top of it. Daniel and Dusty stood on the covered front porch watching them.

"Welcome!" said Daniel, as Dusty was trying to make sense of the new guests' scents.

"Hi, Daniel, nice to meet you!" Lauren and Antwan both said with a slight time delay between them, as if they were harmonizing.

"Well, come on inside. I'll give you a tour."

The three went up the wooden stairs of the porch and into the foyer, which led to an open space and a living room filled with dark leather furniture and a large wood-burning granite fireplace.

Daniel's bedroom was on the main floor, along with a bathroom and an open kitchen that connected to the great room. The guest room upstairs was Lauren's bedroom and bathroom, with a loft area and a balcony with a wide view deeper into the property, where Daniel's observatory prominently stood on top of a secondary mountain summit.

The east-facing radio telescope dish was positioned along the trail a few yards below the summit. The west-facing dish was approximately the same distance from the observation dome on the other side.

Daniel escorted them outside to a stone pathway up the mountain to his white-domed observatory. Inside, a refractor telescope was attached to a sophisticated motor driven mount, bolted in to the cement in the center of the space. Around the mount was a curved desk custom fit for the space, housing several curved monitors and a laptop. A cabinet underneath contained the brains of the operation. A row of computer towers operated the roof motors, telescope mount, cameras, radio telescopes, and control programs.

"It must take a lot of memory and processing to collect, store, and interpret all of the data from your telescopes," Lauren noted.

"Yes. I have mirrored data storage as well," said Daniel, vaguely gesturing to another smaller cabinet.

"This is amazing," said Antwan, trying to soak it all in. He was looking at the control software on the monitors, the placement of the radio telescopes, and the mounted telescopes.

Antwan strapped his freezer transceiver behind Daniel's Polaris Ranger and slowly drove it up the path to the observatory. *I'd better not mess this thing up,* he thought, *we've got to be able to use it tomorrow.*

With a capable receiver and transmitter, the three took this opportunity to continue experimenting with the frequency, taking their minds off the stress of the impending presentation.

Thursday morning, Antwan awoke to the smell of bacon. The sun had yet to peer above the mountains to the east, the San Francisco peaks just north of Flagstaff where they would be heading this day. He thought Lauren was still asleep in the bedroom since the door was shut. Then he heard their voices downstairs. He was surprised to see Daniel in a suit and tie, a stark contrast to his normal casual look in jeans and a T-shirt. *This is probably what he wore every day as a professor,* Daniel thought. Lauren had dressed for the occasion as well. She wore a business-style black dress with heels in her hands and flats on her feet. Fortunately, he had also brought a suit.

After breakfast, Daniel helped Antwan load up the freezer. Daniel started his truck, and Lauren jumped in the front seat and said, "Let's make history." Then they started down the driveway, following Daniel's

RAM 2500 on 10 miles of dirt road then about an hour on the interstate until they reached Flagstaff.

Antwan was glad Daniel had lent him a tarp. The chest freezer was too tall to close the bed cover, and it would have been covered in dust from the roads after following a truck. He tried to keep some distance from Daniel, but he didn't want to lose him either.

They arrived in Flagstaff, then navigated through the streets of the mountain college town. *This would be a great place to go to college,* thought Antwan, comparing his experience in Los Angeles to the small town of Flagstaff, roads lined with pine trees, and an ever-present mountain peak peering over the town from the north. It sure was a long way to the beach.

They continued down a residential road lined with houses with parked cars and trash cans along the sides of the road. Antwan thought that maybe they had missed a turn, until they passed a city park and he saw a brown sign pointing up toward the observatory. They ascended a tree-lined road up a hill until he saw a white building through the pines, a dome that was the first indication they were close to their destination. The road curved sharply to the right, and they passed through the rock markers indicating they had reached the Lowell Observatory.

Chapter 10
The TED Talk

The science and technology TED-style conference was being held in the Lowell Universe Theater in the Marley Foundation Astronomy Discovery Center. The space was impressive, featuring a wraparound screen and a ceiling screen, like a planetarium, but with a stage for presenting science content. The room could fit around 180 people, and when they arrived, it was apparent that today's conference would be at or near capacity.

After checking in to the conference, Antwan called Pedro to make sure everything was set, while Lauren and Daniel wheeled the freezer into the building and around to the side of the stage. Daniel set up a video meeting with Pedro and a live stream of his own observatory. He had set up his radio telescopes to monitor the frequency before they had left. For the demonstration, there would be two receivers at all times and three radios: one at his property, one on the beach in Oceanside, and one in the room with them.

After an interesting lecture on the discovery of the latest exoplanets and an update on the NASA Artemis mission, Daniel, Antwan, Lauren, and Pedro were up.

Antwan hit record on the meeting and sent a text to Pedro to log in to the meeting. Daniel, or Dr. Sullivan here, started with introductions, including education credentials that supported the credibility of the presenters:

"Good morning. My name is Dr. Daniel Sullivan, retired professor of physics from the Polytechnic Institute of New York University. With me today are Antwan Richard, B.S. electrical engineering, University of Southern California, and Lauren McCartney, B.S. oceanic and atmospheric sciences and M.S. geophysics, University of California San Diego."

"As described in the itinerary today, we are here to share a recently discovered 'radio frequency phenomenon.' Don't tune out quite yet, though. The implications of this discovery could be astounding," Dr. Sullivan said.

"At this point, I'll turn the mic over to Mr. Richard and Ms. McCartney."

"Thank you, Dr. Sullivan," said Lauren, as Antwan wheeled the transceiver onto the stage and broadcast the video chat and the control screen of the transceiver to the massive, curved screen on the wall behind them. Lauren described the build of the transceivers, glossed over some of the complications of the build and how they were solved, and then started to explain the testing they had done and what they had discovered at their frequency.

This was the perfect time for Antwan to take over and explain what was about to transpire.

"Hello, everyone. Today, I have my associate, Pedro, on the screen. Say hello and show them around."

"Hello, everyone! I am Pedro Cruz, here in Oceanside, California, at my surf shop. The next time you are in town, please stop on by!" He had shamelessly plugged his store as he panned the camera toward the beach and back to the twin of the freezer on the stage.

"Thank you, Pedro. Now we going to start by transmitting a continuous wave signal from the beach at Oceanside, California, at 1.0-watt power. Go ahead, Pedro."

The indicated signal strength on the control screen popped up to exactly 1.0 watt, as the built-in speakers let out a steady tone.

"Now, we are going to adjust this frequency up one times 10 to the negative 24th of a Hertz, match the two, then back down through the previous frequency, and down one more time."

The tone and signal disappeared as Pedro shifted the frequency up. Antwan shifted his up with no change, then brought it back down. Next, Pedro dropped the frequency, returning the tone and signal, and then dropped it once more, causing it to disappear. Antwan matched it with no success, as expected, then raised it back up.

"Alright, this time we're going to play some music," said Antwan. Pedro interfaced the radio with a Bluetooth adapter, then pulled up Beethoven's C-sharp Minor Sonata, more affectionately called "Moonlight Sonata," and began broadcasting on AM. Antwan had chosen this classic because it was in the public domain. YouTube and other online video services would not have a copyright infringement reason to block audio from the presentation video they were making.

The Lowell Universe Theater was suddenly filled with the gradually increasing melody. Antwan muted the speakers from his transceiver and, selecting the audio from the stream at Daniel's observatory, verified the music was playing there, too.

Now unmuted, Antwan removed the antenna from his device. The music kept playing.

"Pedro, please remove the antenna from the transceiver." The music kept playing.

Antwan instructed Pedro to adjust the output power a few times, just enough to show the audience that the power output matched the signal reception strength both on his transceiver, but also on the reception of Daniel's radio telescopes about 60 miles due west.

"Thank you, Pedro. I appreciate your help. Now I'll turn it back over to Dr. Sullivan."

Dr. Sullivan began reading his prepared speech to the audience:

"In early physics, a preeminent theory once held that light traveled through a medium called luminiferous ether. As our knowledge evolved, we discovered that light held properties of both matter and energy and that ether did not exist.

Outside of our atmosphere, everything travels through the void of space, but then we found that even the void itself is not empty. The void of space is full of dark energy, which is attributed to be the cause of the expansion of space and the universe.

Today, we present to you a new discovery that could challenge what we understand in modern-day physics.

Imagine a frequency that appears everywhere all at once with no time delay.

A god frequency omnipotent with no loss in power regardless of distance. An energy that is inert to interaction with matter, like a neutrino, but taking it a step further, inert to interaction with dark energy as well.

At this specific frequency, for some reason yet to be discovered, energy seems to only react with its own frequency.

In trying to understand this phenomenon, it appears evident that the speed of travel of this energy is immeasurable. Faster than the speed of light.

Quite possibly, this energy wave could be transmitted from the far reaches of the galaxy, the universe, or beyond, and be instantaneously received right here on Earth. We could potentially communicate in real time with someone or something literally light years away.

With this new perspective, I propose a new theory of the speed of light that has been a nearly universal constant in physics for centuries:

Imagine if you will, a skydiver in a freefall. The mass of the skydiver interacts with the gravitational acceleration of the earth and pulls the skydiver downward. Then the air resistance pushes upward causing an equilibrium constant speed of approximately 120 miles per hour. This is called terminal velocity.

Now consider that light is the skydiver and dark energy is the air resistance. This is what gives us the value of the speed of light in a vacuum.

Remember that the void of space isn't a void at all. It consists of dark energy. What we have discovered is an energy wave at a specific frequency that is not held back by the same limitations of matter or light. It is not held back by – immune to – the resistive effects of the medium of the universe: dark energy.

We hope to continue researching this radio frequency phenomenon by exploring partnerships with education and research entities.

Thank you very much."

Chapter 11

The Hype

And that was it.

After a few more 20-minute presentations, the conference came to an end, with a few astonished attendees, but none who were immediately offering aid for what comes next for them.

Even knowing the conference was live streamed and available later, before leaving Wi-Fi for at least eight hours, Antwan took this opportunity to upload their presentation to YouTube. *Perhaps some deep pockets caught wind of it and want to support out research,* Antwan thought. *I mean, who knows what could happen?*

Dr. Sullivan helped load up the transceiver, and Antwan gave him the tarp as they parted ways. Antwan and Lauren were taking the southern route through Phoenix to get back to San Diego rather than drive through L.A. again.

The drive out of the mountains was an easy one. The stress of the past few days had been lifted, and there was a general sense of relief for both Lauren and Antwan, a feeling that life would slow for a little while and perhaps come back to a new normal.

The timing of the sunset was on point as they pulled into the Sunset Point rest area. Already behind the Bradshaw Mountains, rays of sunlight illuminated the underside of the clouds in the sky, with various shades of pink and orange. Antwan and Lauren made it to the scenic overlook at the perfect time, and Lauren insisted they take a selfie to capture the moment. She pulled him in close and it seemed the most appropriate time to both of them for a first kiss.

The 80-degree weather of Flagstaff during the day gave way to the 100-degree weather of the Sonoran Desert night. Antwan closed the moonroof and turned on the AC for the first time today. On the flip side of Phoenix, the city lights gave way to the pitch darkness, interrupted by the occasional brightness of a prison, town, or farm. A prescription for sleep, and as Lauren drifted off Antwan drove through the night to get them back to their Pacific coast. Antwan pulled into his parents' driveway as the sky was starting to light up and parked next to Lauren's BMW.

The next few days were pandemonium.

The conference video had gone viral. YouTube lit up with references to God Frequency and the hashtag #godfrequency was trending all over social media. TikTokers, YouTubers, news networks, science talk shows, bloggers, and science and technology websites posted articles pontificating on the implications of the discovery. Conspiracy theorists on radios, podcasts, and websites were going unhinged. Clickbait articles were already picking up on the trend to suck people in by posting articles on the topic.

It's amazing how quickly a phone number can get out, thought Antwan. By Friday afternoon, his phone had been ringing once or twice an hour, then increased in frequency. By midday Saturday, the calling was non-stop. Unidentified phone numbers from all over the country, as well as Caller ID verified phone numbers from a wide variety of media outlets filled up his phone log.

Friday afternoon, with building media attention, Antwan sent a group text to Lauren and Daniel requesting they discuss their next steps. They agreed to meet via video chat on Saturday at 11:00 in the morning. Daniel sent out a meeting invite via text. By the time Saturday rolled around, Antwan had to scroll down nearly 50 texts to find the one with the meeting invite. *So much for the auto spam blocker,* he thought.

When Antwan logged in to the meeting, David was waiting on his balcony, with the topside view of junipers and the white spec of his observatory behind him. Lauren logged at nearly the same time on her back terrace, with the pool umbrella and a clear sunny sky behind her, along with the occasional glimpse of a plane towing a banner in the distance. Antwan didn't have a comparable scenic backdrop as his companions. He was in his workshop, but he was comfortably in private and not concerned about media interruptions at the moment. He had already placed his phone on airplane mode.

"Well, this is what we wanted, isn't it?" said Daniel.

"This could be the means, but what I'm hoping for is the ends," said Antwan.

"I think this is the best outcome we could have hoped for," said Lauren. "But we should have had a strategy for how to proceed."

"I don't know if you can strategize going viral," Antwan countered. "It won't take long before people are at our door, Lauren. It's impossible to not get doxed when you have a ham radio license," referring to the Federal Communications Commission (FCC) requirement that all valid amateur radio licensees maintain a valid name and address associated with their assigned callsigns in a publicly available database.

"This is your 15 minutes of fame," Daniel implored, minimizing his own association. "I suggest you use it to inform the public of your discovery and hopefully get interest from the scientific community to fund further research."

"We need to decide who to talk to and when. Do we combine efforts or split up?" Lauren asked. "What about you, Dr. Sullivan?"

"I've no intention of travelling to studios, but I can participate remotely," he replied.

"I'm game. Let's do this," said Antwan.

"Okay, I'll compile a list and we'll prioritize and reach out," said Lauren.

"We also need a list of ones to stay away from," Antwan replied.

"Before we go any further," Dr. Sullivan added, "We cannot divulge the frequency. To make it public would be to share it with anyone who has the means or the knowledge to access it. This is an international security issue."

"Agreed," Antwan and Lauren said simultaneously.

Chapter 12

The Deception

Agent Devereux breathed a sigh of relief and took off her headphones. *At least we won't have to worry about it voluntarily getting out,* she thought, as she closed the video chat window. It's unfortunate what comes next though.

This case was a divergence from the typical cybercrimes Levy Devereux had been assigned to the last three years as a federal agent. Most involved cyber forensics, typically for cryptocurrency. She had been successful at tracking down stolen crypto equivalent to billions of dollars in U.S. currency.

Due to her success, her colleagues stopped referring to her as Blanche behind her back, an ironic nickname, given that aside from the body type—a typical misogynistic fixation—she was very different than the *Golden Girls* character.

She was relatively young, married to a woman, and had a beautiful daughter in grade school. They did, subsequently, begin referring to her by her actual and uniquely appropriate first name, Levy, which *Merriam-Webster* defines as to impose or collect by legal authority. This case she was assigned was a step up, even if was not a crime. As her targets had stated themselves, this was an international security issue.

There were some similarities to past cases, particularly the confiscation of electronics, this time as part of an effort to keep the technology and, most importantly, the specific frequency of the scientific anomaly, out of the hands of bad actors.

She had already infiltrated Antwan's and Daniel's personal email accounts and surgically removed the attachment containing the frequency. That didn't mean just deleted. She removed it and passed it on to other government agencies, highly classified, of course.

Her team was monitoring phone calls and texts of all three of her targets since midday Friday. By Saturday, she already had to request additional resources to track the origins of all of them.

She was not intent to chase down Antwan like the kid in the movie *Mercury Rising*. A closer analogy would be *Oppenheimer*, where he developed the technology for the nuclear bomb. Now, just like the bomb, there will be an international race to develop the technology and find the frequency. And, just like the bomb, there needs to be an international group to regulate the use of it. But that is somebody else's problem, and that takes time. It was her job to give them that time.

Agent Devereux was tasked, specifically, with keeping the frequency and the technology to access it from being made public. This was surely a temporary effort until other agencies get ahead of this. But with all eyes on the god frequency story, she would have to tread lightly on how she accomplished the task.

The Federal Trade Commission (FTC) website was easy enough to access, but the number of private databases listing amateur radio call sign data was quite surprising. Still, she managed to corrupt any online callsign listings for Antwan and Lauren. *This will keep the media from crowding their driveways for just long enough,* she thought.

Next, she would have to lure them away, so media attention was elsewhere for her moment. Confiscations go much smoother when the targets are absent. She knew their motivation and had access to significant government resources to draw them out. The choice at hand was which string to pull: NASA, DARPA, NSF, DOD, DOE, ONR, JPL… or perhaps even a non-profit like SETI. National Science Foundation would be the obvious choice for funding but a boring choice for travel. She had it! The United States Naval Observatory (USNO) is the official source of time for the U.S. Department of Defense, and a standard of time for the United States.

Levy Devereux wasn't dumb. She had graduated Quantico, but her real credentials came before her career, earning a PhD in algorithms, combinatorics, and optimization from Georgia Tech before she even knew what to do with it. She recognized the potential significance of the god frequency, if it was, in fact, real. And she knew that there wasn't a better way to measure something greater than the speed of light without the most accurate time standard available and that was the USNO.

The perfect carrot.

A few phone calls later and all was arranged. Emails were sent to the two of the three targets, knowing fully well that the professor would likely not take the bait. He was the least of her worries. His property was so remote, it is very unlikely the media was camping on his porch. Lauren would take the bait. The USNO was like Mecca for oceanography, and if Lauren went, Levy knew Antwan would accompany her.

Douglas Hemme

Subject: USNO Research Invitation

Ms. Lauren McCartney & Mr. Antwan Richard:

Your presence is requested at the United States Naval Observatory, Washington D.C., for a special opportunity to participate in unspecified testing of a recently discovered frequency anomaly.

Transportation has been arranged. Please be prepared to depart your residence at 0500 PDT Sunday, June 29th.

Sincerely,

Rear Admiral
Commander, Naval Meteorology and Oceanography Command

By early Saturday afternoon, Lauren had compiled, categorized, and prioritized a list of media outlets that had reached out to her and Antwan so far. Antwan had taken the liberty to respond to some news service interview questions sent via email. This was by far his preferred method of interaction. Lauren, on the other hand, had granted a handful of phone interviews.

Both were lucky to have avoided the troublemakers vying to abuse their newfound popularity to their own ends. Calvin Hodges seemed to be the loudest voice rising from the cacophony of deniers and conspiracy theorists, throwing out terms like hoax, blasphemers, liars and heathens to boost followers and gain popularity through the denigration of others. Lauren, Antwan, and Dr Sullivan were just the latest targets of the anti-science crowd. They would surely become too boring a target before long and the collective gaze of contempt would move on to the next popular scientific achievement. But until then, they did well to avoid the crowd.

They were scheduled to do a night show the following week, but a local San Diego news station and newspaper were bringing crews to Antwan's workshop to do a piece on his transceiver build and discuss the god frequency, a name that seems to have stuck following the viral popularity of Dr. Sullivan's monologue. He would have suggested a different term to avoid the wrath of the sensitive culture warriors.

Only a handful of news outlets had outright asked for the frequency value, purely for entertainment's sake. Both Lauren and Antwan easily sidestepped the issue. They had agreed to accept the invitation to the USNO, as if they were given an option in the request.

Promptly at 5:00 a.m., a black SUV with government license plates pulled into Antwan's parents' driveway. He had packed a backpack with a change of clothes just in case. There were no details in the invitation about the itinerary of this trip.

Antwan walked out of the house and down the walkway as a man in a suit opened the passenger door for him to enter. No words were exchanged, which suited him just fine. From there, they were headed down "'the 5'" and began the familiar route to Lauren's parents' home. She, too, was waiting anxiously and walked out as soon as the SUV came up the long, narrow drive. In less than a half hour, the two were on a C-40A Clipper aircraft taking off from NAS North Island. The C-40A was the United States Navy version of the Boeing 737-700 for high-priority cargo and passenger transport.

Antwan wasn't sure, according to the Navy, which one they were.

The plane was empty, except for two military shipping containers loaded in the rear of the fuselage, the one suited passenger from the SUV, and a woman in a business suit who walked back from the cockpit and introduced herself as they boarded.

"Hi, Lauren, Antwan. My name is Levy Devereux. I'll be escorting you two to the United States Naval Observatory."

Chapter 13

The Confiscation

Back at Byron and Maria's home, two SUVs and two box trucks pulled up the driveway. One box truck backed up to Antwan's workshop, and a dozen agents exited from the caravan. Two agents waited near the front porch of the home with a freshly inked search warrant. Both chest freezers were promptly loaded into military-style cargo containers, then loaded onto a box truck. The roll-up door was closed, and the truck descended the drive.

The other agents entered Antwan's workshop and began collecting documents, computers, and radio parts, loading everything that could possibly contain information about the god frequency. All of Antwan's amateur radios were carefully loaded into shipping containers and placed into the second box truck.

Byron woke up as he normally did around 5:30 a.m., even on the weekends, and went to his bathroom to begin his morning routine. Outside his window, he heard the commotion outside. *What the hell is Antwan doing out there?* he wondered, thinking perhaps he had gotten lost in another project that kept him working through the night. It became quickly evident that something more nefarious was going on. He checked the time on his cell phone and noticed he had no cellular service, and his Wi-Fi would not connect either.

Byron put some clothes on, tucked his firearm into his waistband behind his back, descended his stairs, and opened the front door. The two neatly dressed agents waiting on his porch intercepted him and quickly identified themselves with badges, handing him the search warrant. Thanks to the Patriot Act, any relevance to potential terrorism streamlined the process of obtaining legal authority to enter residences and seize property. In this case, the rationale given to the signing judge in the secret court was that it could be a problem if a terrorist obtained the god frequency and the technology to access it while unsecured in Antwan's workshop.

In the warrant review, Levy proposed that if a foreign adversary or terrorist were to gain access to the god frequency, with no transmission power losses, it would not take much amplification to overload and destroy any other receiver actively listening to the frequency. Alternately, they could broadcast a steady signal on the frequency, effectively obfuscating any potential uses of the frequency with no currently known method to identify the origin of the signal. The judge was not swayed by the second argument but talk of universal destruction by bad actors and foreign adversaries was usually a recipe for guaranteed approval. The warrant was approved as submitted.

Byron read the papers and understood he had to comply, then pulled out his cell phone to inform Antwan. Still no service. Still no Wi-Fi. Very odd since he normally had a very strong signal from his home. He sent a text message, intended to reach Antwan once his service had returned.

About an hour passed and the yard full of activity was beginning to be illuminated with natural light as the last items were loaded into the back of the box truck and the agents got into their vehicles and departed. Byron watched them drive down the street and turn out of the neighborhood when his phone began to buzz repeatedly with multiple notifications that had been stored in the various purgatories of the internet until service was restored to the end user. His phone service and Wi-Fi had been restored. He knew the timing was no coincidence.

Lauren's parents were not home when the agents arrived. By then, Lauren and Antwan were cruising at 36,000 feet above the Mojave Dessert. Two agents walked to the end of the long, narrow driveway and waited. The other agents quickly got to work, picking the lock, disabling the alarm system, and searching the home. Lauren's radio setup was located in a room near the back of the property. The computer and radios were carefully removed and placed into shipping containers, then loaded into a box truck. Less than half an hour later, the house was locked up, alarm reactivated, and agents gone.

In the early afternoon, a convoy of three Humvees arrived at Dr. Sullivan's gate, they waited patiently, aware they were being watched, until Daniel came walking down the winding drive with Dusty by his side. He was not completely surprised by their arrival. They had already identified the international implications of the discovery and he had advertised it to the world. Still, he was compelled to review the paperwork before allowing free rein on his property.

Once satisfied at the authenticity and credibility of the agents, he opened the gate and stepped aside as the Humvees continued to the house and up the walkway to his observatory. Daniel, more concerned for his younger friends, pulled out his phone to call them, but he did not have any cellular service or Wi-Fi connection. Clearly, at least one of the Humvees had an active signal jammer inside.

Daniel watched as the agents carefully removed each one of his computer towers from the cabinets inside his observatory, placed them into a shipping container, then wheeled it down and loaded it into the back of one of the Humvees. It was going to take a lot of time to re-install the operating programs for his equipment onto new computers, he lamented to himself. Replacements weren't going to be cheap either. *Perhaps,* he thought, *I'll take this opportunity to upgrade to server instead.*

When the agents were satisfied the house and observatory were adequately searched, they handed Daniel an inventory of all items seized and departed, single file down the driveway and toward the highway. The lead agent, Jose Burton, informed Daniel "You can no longer discuss this frequency anomaly, or anything else associated with the frequency anomaly, with family or friends or colleagues. Fail to follow these guidelines and you will be subject to prosecution up to and including imprisonment."

Daniel nodded in acknowledgement.

Daniel sat on his balcony drinking his coffee, hoping his friends were okay. With his phone service restored, he tried calling Antwan, then Lauren, but only reached voicemails, and he knew better than to leave one. He sent a group text message that went unanswered,

Back in his bunker, Daniel keyed the frequency into the controller and once again had ears on the god frequency, and with no need to reposition the dishes to tip his hand. He then locked the notebook in his bunker's gun safe.

Daniel wondered how long it would take the federal agents to figure out that his receiver precision originated from the hardware in the receiver rooms beneath his radio telescope dishes—not in the control applications of his computers.

Chapter 14

The Persian

The buzzing sound of the drone's propellers was drowned out by the light breeze and rustling of palm fronds above the mostly still Oceanside neighborhood, but the flurry of activity continued in the front yard of the Richards' residence. Box trucks and SUVs crowded the driveway and men were coming in and out of the workshop. *It is too late for a simple breaking and entering job,* thought Ramin.

He was content with his job as a computer repair technician working near Naval Base San Diego. He had an added responsibility to screen the files and email accounts of clients for information of ships' movements, readiness, or capabilities and feed that information back home. He had a computer science background. He wasn't properly trained for burglary.

The message he had long feared, given the rising political tensions between countries, had just arrived, and the task he was given was not what he expected. Steal a radio inside a chest freezer! They've got to be kidding. So here he was, early hours of the morning, scoping out the residence, but it was too late.

He brought the drone back over to the shopping center parking lot, where the nightshift employees were just finishing up restocking the shelves of the stores and coming out to go home. He landed his drone quickly, got in his car without bothering to fold it and pack it up, then departed. He didn't want to answer any questions.

Ramin headed north on California Highway 76 to Interstate 15 in his Toyota Corolla. He was expected to waste no time in implementing plan B. *A number,* he thought, *how hard would it be to get a number from an old man?* He had the day off of work, but he would likely need to take Monday off, as well. *Going to visit a sick relative is a good enough excuse,* he thought as he made the phone call to his work's voicemail. He was somewhat optimistic he could return to his life in San Diego, regardless of his activation.

Passing through a border patrol checkpoint, he was questioned thoroughly as usual. "I appreciate the work you do," he told the officer in perfect English, "everyone should come here the right way," he volunteered as his identification was scrutinized. He really believed the statement. Everyone should go through the proper channels to enter the country, even if their motives are disingenuous. *Especially if their motives are disingenuous,* he thought.

Ramin would have liked to have continued his current life indefinitely, masquerading as an immigrant who sought and received approval for asylum, but duty called. Now he was leaving the coast, which reminded him most of his childhood town. Jask was a port town and, like San Diego, his government built a naval base there when he was young, redefining it as a strategic location. From there, his country's navy could block the Strait of Hormuz and ultimately entry into the Persian Gulf. He was proud of his town, and likewise proud of the work he did for his country, even if it took him out of his comfort zone.

By evening, Ramin found himself travelling up a dusty road, just off of Interstate 40. He had traveled the long way around the mountain where he was going, intentionally avoiding the obvious approach where he might be seen. This was one of the few tactics he retained from his brief training before being sent abroad. He then found a clearing and pulled off the road and through the trees. Cows were meandering through the property, but he had seen a State Trust land sign, so he assumed there wouldn't be anyone coming to round them up. He placed a fresh battery in his drone and fired it up.

Rising over the clearing, he used the gimble to angle the camera toward the mountain. Up to 150 meters, the sound was once again minimal from the ground, he then began its journey. In the trees below, Ramin spotted a faint outline of a vehicle rooftop. Zooming in, it appeared to be a work van of some sort, and not as dilapidated or abandoned as he hoped. *A possible complication I will have to deal with,* he thought.

The LED lights of the drone had been carefully disconnected so it could not be seen easily, but with the dim remainder of sunlight behind the mountain, his only reference was the dim lights coming from a lone cabin on the side of the mountain. He found his mark.

After the drone returned, Ramin carefully folded and packed it up, opened his trunk, pulled out the foam tool tray and removed a 3D-printed plastic pistol form-fitted into the underside of the tray. He put the gun in his waistband behind his back, grabbed the tire iron, and set out toward the van and the property.

Chapter 15

The Conspiracist

Following the now infamous god frequency lecture, as it since became known, a massive number of comments flooded the internet, email inboxes, video and social media comments, and threads. Among the average user comments were more fringe ones, like "how dare you use God's name in vain," or "we should not be interfering with God's work." But the most concerning was the consistent condemnations of their discovery on a popular underground webcast by a man who went by Calvin Hodges.

Claiming a public persona as a religious fundamentalist, Calvin developed a following by sharing unfounded claims against anyone that he perceived as ideologically different. He found that the more shocking or ridiculous the claim, the more activity he would generate, and the more ad revenue he would receive.

Sales of his persona merchandise also trended directly with the number of viewers he drew in to his podcast.

The popularity of the god frequency story, combined with its inherently charged name, was a perfect storm for making money and increasing his own popularity. For three days straight, Calvin replayed the presentation video on his podcast, railing against everything from deviation from traditional gender roles to immigration, denouncing interracial couples with racist undertones. He attributed the economic problems of California to all of these things, and since they were educated, he was able to attribute blame to the culture at universities and academia, writ large.

He didn't stop there. He wanted the four of them—Antwan, Lauren, Daniel, and Pedro—to suffer consequences for the comparison of their discovery to God. He angrily charged that they shamelessly blasphemed and needed to repent. He also claimed that the discovery was a hoax and that the radios were simply connected to each other via the internet.

Calvin charged his followers to find Antwan, Lauren, and Dr. Sullivan, and dox them. He had crafted, through his podcasts, a talent for thinly veiled threats that didn't ascend the threshold of legal liability for his language, but effectively cultivated the emotional reaction of his followers, leading to a situation that was highly volatile for his targets. He did not get his hands dirty, but cultivated an environment where people could channel blame for their frustrations in life and transfer it to another person or group using their existing prejudices without even being privy to the manipulation.

Antwan and Lauren were not as private on the internet as Daniel. Their parents' houses were found rather easily, but because they were in populated and well-policed areas in California, they were relatively safe from Calvin's flock. Daniel, on the other hand, lived in an area more likely to follow Calvin, more likely to act on his direction, and less protected by populous, geography, or law enforcement.

Internet sleuths, using information gathered from their viral presentation, soon identified Daniel's radio telescopes on satellite map imagery. Despite the lengths he took to keep his privacy, they would no longer keep him safe. His coordinates were shared and spread like wildfire through fringe bulletins and threads on websites like 4Chan and Reddit. It was only a matter of time before Calvin's hate translated to someone else's actions

Leonard Young was one of Calvin's most avid followers. He was a product of a strict religious upbringing and was disowned by his family when his girlfriend became pregnant. Regardless, knowing no better family dynamic, he attempted to translate his strict traditions into his own family. They married and his wife quit her job after their daughter was born. At that point, he controlled all facets of their life, as he drove them to traditional family roles. However, Leonard could not keep a consistent job for more than a month and it weighed on him.

After another night of heavy drinking, which wasn't unusual, his wife criticized his financial priorities and how he sought to physically reinforce the roles in the family. Luckily for her, he'd had too much to drink and passed out.

The next day, he was alone in their apartment. In a haze in this period of his life, the next thing he remembered was being evicted. Now the 23-year-old was navigating a homeless lifestyle as best he could, living out of his Ford Econoline van, spending the winter months in Phoenix, then driving upstate to northern Arizona for the summer. His routine was not unlike the "Snowbirds" who move to Phoenix from Canada or the northeastern U.S. during the spring and winter months, then back to their normal life for the summer and fall. But he wasn't part of their "godless elite pushing an atheist agenda and forcing indoctrination into their cabal through colleges and universities."

No, Leonard felt like he was pushed out of the workforce by refusing to be extorted into higher education. He was a victim of religious persecution and unenforced immigration laws that discriminate against hard-working United States citizens, and when Calvin

Hodges spoke, Leonard felt like it was directed straight to him.

He accessed free Wi-Fi from the parking lot of the nearest McDonald's, any coffee shop with free internet, or when all other methods were exhausted, the library to access Calvin Hodge's podcasts. Leonard didn't miss a single one. Even a job wouldn't have kept him away from it.

If Calvin asked for a sacrifice, Leonard would make one of himself for the common good. And Calvin made one request, one call to action, that resonated with Leonard: Make Dr. Sullivan pay for taking the Lord's name in vain and lying about a discovery that challenges the laws of the physical world, which only God can do.

Leonard set about finding Dr. Sullivan to teach him not to mess with God, or his followers will make him suffer the consequences. It didn't take him long to find the coordinates for Daniel's property.

Leonard sat in the driver's seat of his van, parked in a clearing, surrounded by juniper trees, somewhere on State Trust land west of Flagstaff. He had the coordinates in the phone's map, but he wasn't great reading the thing. Still, he knew he was close. All he had to do was wait until the sun went down and he would find what he was looking for: a property with electricity on the side of a mountain with large satellite dishes. Someone else had done all the hard work finding it. Now he just had to wait.

He put his revolver into the right pocket of his paint splattered jeans, tipped his trucker hat forward over his face, and drifted off to sleep.

Chapter 16

The Confirmation

The C-40A Clipper landed at Andrews Air Force Base in the early afternoon, Eastern Time. The five-hour flight did not contain the luxuries of a commercial flight, but they were provided some small snacks, coffee and water bottles. Coffee was a staple in the military. Byron still ritualistically made a pot of coffee every morning before the sun came up, pouring the bulk into a green Stanley vacuum-sealed container. No milk, no sugar, no creamer. Just black. It would be empty by mid-morning.

The plane taxied off the runway where a 2.5-ton light medium tactical vehicle (LMTV) and a black SUV with dark tinted windows were waiting. Levy escorted them to the SUV while her quiet, well-dressed companion stayed behind, presumably to oversee the unloading and transfer of whatever was contained in the shipping containers in the back of the plane.

Leaving the Air Force base, Levy instructed the driver to stop for food. The driver pulled down a two-way street, lined with parked cars near Nationals Park, and dropped them off at a restaurant called Mission Navy Yard, serving Mexican food and known for its weekend brunch fare. The two of them felt like they were receiving the VIP treatment, with the escort, plane ride, and food at restaurants. Lauren had an underlying feeling that something wasn't above board, but she did not want to ruin the experience by sharing her suspicions with Antwan. He, on the other hand, had been too preoccupied with the experience to even turn his phone back on after they landed.

By late afternoon, the made their way to the northwest side of the Capital City, passing by the British Embassy and the Finnish Embassy, and entering the United States Naval Observatory from the North Gate. Agent Devereux spoke with the guards at the gate and received two visitor badges. Lauren thought it very peculiar that they were pre-printed with their names and pictures on them. Her experience with federal government bureaucracy was not nearly as efficient as the example being set today.

The vehicle meandered through the complex, passing by the United States Vice President's residence at Number One Observatory Circle. Originally built for the superintendent of the Naval Observatory in 1893, it was taken over by the Chief of Naval Operations, then eventually every vice president from Walter Mondale to present.

Finally, cresting the hill, their vehicle stopped at the grand main entrance of the U.S. Naval Observatory's Georgetown Heights Headquarters. Agent Devereux escorted the pair into the James Melville Gillis Library, a massive circular room with two stories of books lining the walls. As much as a museum as a library, the room houses over 90,000 books, including a substantial collection of 19th century works and an 800-volume rare book collection.

The grounds of the USNO house were immaculate, befitting its place in this Georgetown neighborhood, its history, and its current role in United

States politics. It was surprising that its important scientific responsibilities continue here, throughout its history, all the way to present day.

Agent Devereux took her leave, allowing them to browse the library and view the exhibits as she departed the room. Antwan, seeking to take pictures in this extraordinary room, turned on his phone. His phone connected service from the nearest cellular tower, and the notifications flooded in. This wasn't a surprise, given the attention they'd been receiving the last few days, so he took little notice at first. He took a few pictures and put his phone back in his pocket to ascend the metal spiral staircase to the second floor. Taking his phone out a second time, he began to dismiss notifications until he came upon the missed calls and text messages from his father, Byron.

"Lauren!" he cried out to her from across the room.

"What's wrong?" she said.

"They took it all. The radios, my computers."

"I had a feeling," she said, as Agent Devereux walked into the room, accompanied by three men, two in suits and one United States Navy Commander.

The Commander introduced himself as the Deputy Superintendent of the U.S. Naval Observatory. The other two introduced themselves as agents, without specifying agency. Antwan wasn't the best at remembering names the first time, but he did notice the intentional omission from the agents. The best he could do is remember the Commander's name reminded him of the actor John Cusack, which he knew from the movie *2012*. Antwan sincerely hoped the plot unfolding would have no more similarities than that.

"Levy," said Antwan, "do you know anything about my radios and computers being taken?"

"We were just about to fill you in, if you'll follow us," she replied.

The two followed the group down a hallway, upstairs, through a badged access door, and into a large conference room already occupied by a diverse group of people, civilian and military. Most of them were probably mid-level employees, judging by the ranks of the U.S. Navy officers in the room, ranging from Lieutenants and Lieutenant Commanders, to the Commander who walked into the room with them. The Commander noticeably assumed control of the operation, whatever it was, as he entered the room.

Agent Devereux and the two other agents ushered them into an adjacent, smaller conference room with windows into the larger one, then closed the door. Levy handed Antwan and Lauren each a manilla envelope.

"As you may have heard, due to the potential national security risk, we were required to confiscate any technologies that may be used to access the frequency anomaly you discovered, along with any medium in which the frequency itself may have been stored. As such, we have taken possession of all your functioning radios and computers," she told Antwan and Lauren.

They sifted through the paperwork, which consisted of nearly identical warrants describing vague details of what was authorized and why, also listed detailed descriptions of the specific items taken from each location.

"In light of this inconvenience, I have been authorized to allow you to participate, in a limited capacity, in the testing that will be performed this evening regarding the speed of transmission on the subject frequency," she said.

At this point, Agent Devereux called in a female Lieutenant from the larger conference room to explain to them the testing that was to take place.

"We have interfaced the transmitter portion from one of your transceivers into the Master Clock that exists here at the United States Naval Observatory in Washington D.C. and interfaced the receiver portion from your other transceiver at Alternate Master Clock Facility at Schriever Space Force Base in Colorado. For the last two hours, we have been calculating the precise time it takes, initiating a signal, for the device to transmit. This is crucial in determining the speed of travel of the transmission.

"For this reason, we could not keep the transceivers intact. At the Alternate Master Clock Facility in Colorado, they have already determined the precise time it takes to display a signal from when it received, based on neighboring frequency transmissions.

"We have a different team determining the exact linear path distance between the transmitter and receiver. This distance depends not only on positional location but also elevation and the curvature of the earth between the two points. At an approximation of 1,600 miles with an approximate speed of light of 186,000 miles per second, we would see a timelapse of approximately eight milliseconds between transmission and reception of an unimpeded energy signal traveling at the speed of light. Prior to the test, our Alternate Master Clock Facility will be synchronized to the Master Clock in Washington D.C. using Two Way Satellite Time Transfer to match the Master Clock to within a billionth of a second, so if there is a transient time, we will see it."

Antwan was truly impressed. The research side of the Department of Defense, especially with its civilian integration into projects, was a side of the military with which he was unfamiliar, even with his experience as a military brat. He was proud that the transceivers he built were being used for this test, indicating that none of the various government branches or research projects had achieved the frequency precision he had.

This was the first time on this trip that he became nervous. *What if it doesn't work?* he thought. *What if the transmission doesn't have a range of 1,600 miles or the parameters are characteristic of a typical energy wave traveling at the speed of light?* He felt a sense of personal responsibility, after all the resources and effort put forth to perform this test. *I guess we'll find out soon,* he thought, not daring to express his doubts out loud.

"Before we continue," said Agent Devereux, "I must inform you that this frequency anomaly, the technology associated with accessing it, and everything you are about to see today is classified and should not leave this facility. You have both been granted temporary security clearances for this. You can no longer discuss this, or anything else associated with the frequency anomaly, with family or friends or colleagues. Failure to follow these guidelines will result in your security clearance being revoked, and you may be subject to prosecution up to and including imprisonment."

Antwan and Lauren, glanced at each other with stoic looks on their faces. Not two days ago, the two had been discussing how best to communicate their discovery to the world, with a constant barrage of media contacts knocking down their door. Now, they could no longer talk to a single one of them.

"Okay," said Antwan reluctantly, as if he were given a reasonable choice.

"Got it," said Lauren, visibly annoyed by the demand.

One of the silent agents opened the conference room door and escorted them to seating along the wall of the larger room, clearly overflow seating for personnel with roles not important enough to sit at the main table. Large screens on the wall displayed satellite orbits, time data, video links to laboratories, and a graphic displaying a peripheral view of the earth, portraying its curvature and elevations, with a dotted line between two points: Washington D.C. and Colorado Springs, Colorado.

Personnel were sitting along the periphery of the long rectangular table, most with laptops and headphones, talking to who knows who. *The only thing certain is the people at opposite end of the conversations are authorized to talk about the frequency,* thought Lauren, facetiously.

For the next hour, Antwan and Lauren sat quietly, watching the operation unfold. They were allowed unescorted access to the building, though their badges only allowed them access to transit the hallway, breakroom, and bathrooms. The atmosphere of the room reminded Antwan of a SpaceX launch control room, equipped with large wall screens and windows along the wall, but he could tell that it was not routine for the U.S. Naval Observatory to conduct real-time operations. A large conference room table with laptops was not as ideal a configuration as rows of control stations with multiple screens.

It became evident all the small conversations around the table were wrapping up and the main event was about to get started. Times from transmit keying to broadcasting a signal had already been determined using neighboring frequencies but would have to be re-tested at the test frequency.

Commander Jackson Curtis nodded to a Lieutenant Commander, who stood up and began a roll call of the groups involved, apparently assigned as the test director. Receivers at various military sites across the country and the world, as well as satellites, were tuned as precisely as they could in an attempt to detect the signal.

The Lieutenant Commander then authorized the test to begin.

They had apparently chosen to do an initial broadcast of a long burst of a continuous wave at 50 watts. They weren't yet interested in even viewing the results until they verified the time it took from initiating the transmitter to transmitting the signal. Once verified to be in line with the previously measured values, the transmit to receive times were compared: zero.

The actual time the transmission was broadcast and the time the signal was received were exactly the same to a billionth of a second.

The test was repeated several times, including iterations without an antenna attached. All the results were the same. Antwan and Lauren looked at each other, feeling vindicated. None of the other receivers set up across the world or the satellites set up to receive had detected a signal at all. Not even during testing at 100 watts.

There was a flurry of activity during and immediately after each test, double and triple checking their indications and results. Most of the scientists, analysts, and career military professionals in the room were perplexed. Sure, this is what they were looking to confirm, but none of them expected it to be real. They were looked on as pariahs when they had first arrived, but there was a new-found sense of reverence in the room for two of them.

Antwan and Lauren were mentally exhausted from all they had experienced the last two days. They wondered if Dr. Sullivan had continuously been inundated by media interest in the god frequency the time they were gone and intended to reach out to him once their flight landed. Antwan felt somewhat guilty, having gone to the U.S. Naval Observatory without an offer to an amateur astronomer who maintained his own observatory. The library alone was a reason to travel to Washington D.C., not to mention everything else they had witnessed. The invitation was not theirs to make, though.

Chapter 17

The Observation

Ever since his observatory control station computers had been seized, Daniel was spending a lot more time in his bunker. He suspected he might receive more unsolicited visitors to his property with ambiguous motives. He wanted nothing to do with journalists, fellow amateur scientists, or tourists who wanted to interject themselves into the latest viral story.

With his security cameras covering most of the perimeter of his property and his cabin displayed on a dedicated monitor in the bunker, he collected fresh food from his refrigerator and began stocking up for a short-term stay at least until the next news cycle suffocated out their discovery with the next viral story. *The public is fickle,* he thought.

After an exhausting morning consisting of one government raid and several survival preparations, Daniel secured the hatch from the inside, laid down to rest on the bunk across from his underground control hub, then drifted to sleep.

Soon, a loud tonal sound startled Daniel from his sleep. Gathering his bearings, he realized he was in his underground shelter, and the circumstances of his situation came flooding back to his consciousness. The sound, though. *Could it be the motion sensors,* he wondered. He checked his perimeter cameras, then the exterior cabin cameras, and all seemed copacetic. Just the afternoon breeze moving through the juniper trees. The batteries were charged, with good voltages also on the inverter. It wasn't that either. Next, he confirmed that the carbon monoxide and smoke detectors were working properly.

Just as he was about to open his gun safe and patrol the property, the tone came in loudly once more. This time, it was obviously broadcasting from his computer speaker at his desk. *The radio telescopes have been found by someone,* he thought.

The sound was coming from the god frequency.

Daniel continued to monitor and record all the signals he received. The first several signals he recorded were 5-second bursts at 50 watts. The tones remained at 5 seconds as the signal strength gradually decreased to 25, 12.5, 6.75, 3.375, 1.688, 0.844, 0.422, 0.211, 0.105 watts, continuing until it could no longer be registered by his equipment. *Someone is testing the limits of their equipment,* Daniel thought, not considering they were testing the speed of the transmission. Surely the United States government did not have long enough to develop the equipment needed to do this. Perhaps they resourced something they had developed, then repurposed it. Still, he doubted that was the case.

The tones began again as 1-second bursts, starting at 50 watts, then increasing to 100, 200, 300, 400, 500, then to 1000 watts.

Someone or something is definitely testing their capabilities, he thought. It couldn't be Antwan's transceiver, even if the government had seized them, because even as precise as it was, it could not broadcast a signal greater than 100 watts.

No, someone else had it, and that was bad.

Chapter 18

The Convergence

Agent Jose Burton had set up shop in the control house of the communications site that loomed over Dr. Sullivan's property. They gained approval from the utility that owns the site for surveillance purposes. He had successfully tapped into the wireless signal of his surveillance cameras, intercepting the signal without impeding its travel to its final destination in the bunker. He had already placed an additional camera of his own at the main ranch entrance to monitor license plates of cars entering and exiting the area.

Spyware placed in the network via the wireless router in the cabin allowed Agent Burton access to Daniel's laptop. It would have been harder to plant the software if Daniel had changed the username and password from "admin" and "password," respectively, but he still would have gotten in.

By early afternoon, he had established monitoring of Daniel's laptop and began monitoring for any hacking attempts on the network. The laptop and email accounts had been thoroughly scrubbed, including the frequency document that had been downloaded, accessed, then deleted.

By early afternoon, Jose had accessed the webcam and microphone of the laptop and was able to ascertain that Daniel again had ears on the frequency anomaly. He communicated this to his boss, Agent Devereux, via a phone call.

"Yes," answered Agent Devereux

"Dr. Sullivan picked up the broadcasts," said Agent Burton.

"I had suspected as much."

"I also identified a suspicious vehicle entering the property area, registered to a Leonard Young. Should I intervene?"

"No, that's not necessary. Just continue to monitor for traffic and keep eyes on it."

"Copy," replied Agent Burton, as he hung up the phone.

Agent Burton pulled up a near real-time satellite view on a large screen at his makeshift workstation in the control house. His eyes on the location would be gone soon as the sun set in a few hours, limited to the surveillance cameras and low-light imagery from the satellites overhead. He didn't have access to infrared video feed from drones like he had access to overseas.

Jose joined the agency after nearly a decade of military service in the Navy Seals. He would have made it two decades to retirement if he had been given a choice. Unfortunately, a well-placed IED his last tour in Iraq ended his military career and took the lives of two of his fellow sailors. Not one to remain on the sidelines, Jose chose to continue his service to country by joining the clandestine services instead.

Agent Burton made his way out of the control house to get a breath of fresh air. The room had a stuffy smell of electronics and dust, without much to look at. Other than what he brought with him, the only contents of the room were a couple of server racks, a fire extinguisher, a file cabinet, and a table with hand sanitizer. His eyes were dry from watching screens for hours, monitoring internet activity, motion activity, and the transit of vehicles to and from the main road into the area. Most were local property owners with the occasional teenagers wanting to get away from prying eyes in the town, but Leonard's presence concerned him.

He looked down into the valley below him, squinting, hoping to see a white spec of the suspect vehicle he had identified 10 miles away at the entrance off of the paved highway. That is when he heard a faint buzzing sound. There was a drone. He recognized the sound but could not see anything in the clouded sky. This wasn't good. Nothing in Leonard's profile indicated that he was technically savvy enough to operate a drone, let alone intelligent enough to surveille a target in the dark.

Agent Burton immediately pulled out his phone and texted Agent Devereux. Something was about to happen here, and he needed assistance.

"There is a small drone, no lights, over the target location. No aerial surveillance. Possible additional persons of interest nearby. Please advise."

"Grey Eagle in the air in 15 minutes," she replied.

Looks like I get my drone after all, he thought.

At the Camp Navajo Army National Guard Depot, 40 miles east, the four agents received their orders. They then sprung to action, exiting the barracks, jumping into a Humvee, and raced to one of several bunkers lining a road on the south side of the base. Emerging from the bunker was a General Atomics MQ-1C Grey Eagle unmanned aerial system (UAS). Lining the UAS up with the empty base roadway, an uneven but paved road, gave it a clear path for takeoff. Before it was in the air, the agents returned to the Humvee and headed north to the base exit.

Agent Burton connected to the Grey Eagle drone, an upgraded version of the MQ-1 Predator drone made exclusively for the Army, and the drone began to accelerate down the wide, empty road, gaining speed, and eventually lift and altitude. The Grey Eagle adjusted its course through Agent Burton's directions, slightly northward and was in route to the target location. Soon he would have the infrared imagery he was accustomed to in field operations.

By the time Leonard woke, he could hardly see through the darkness into the trees surrounding his van. The clouds obscured any moonlight from aiding in a night hike. He stumbled out of his seat and fell into the dirt, got up, dusted himself off, and went to the back of the van to relieve himself. He recovered his revolver from the floorboard where it had fallen. He opened the back doors of the van and removed a pair of bolt cutters, suspecting he would have a privacy fence to cut through to gain access to Dr. Sullivan's property. He then started to make his way through the woods without the aid of artificial light because he did not want to be seen coming.

Ramin slowly approached the van, pausing to listen and watch for any signs of activity. He was not about to be thwarted from his objective by some local backwoods drunk. When it became evident that there was no one nearby, he inspected the exterior of the van.

The rear was littered with religious bumper stickers, anti-abortion, anti-LGBT, anti-immigration, and pro-Calvin Hodges.

There was a wet spot on the ground, just behind the double doors at the rear. He knew there was someone out in the woods and definitely not his given target. *This could complicate things,* thought Ramin, as he began making his way through the dark, shadowless trees, attempting as best he could to find footprints on the ground. *I was not trained for this kind of work,* he lamented one more time as he quietly made his way towards his targets, listening for any movement along the way.

Tracking the interstate, cruising just below the low layer cumulus and middle layer nimbostratus cloud cover, Agent Burton piloted the Grey Eagle advancing rapidly to intercept its targets. The time to intercept was 15 minutes. Below on Interstate 40, a quick reaction force of four agents was advancing toward the same location. Their time to intercept was 45 minutes. In the woods approaching Daniel's property, two intruders

153

were advancing toward their targets. Time to intercept, unknown.

Soon, there would be a convergence on Dr. Sullivan's property.

Chapter 19

The Confrontation

It had been a couple of hours since the last transmission was received on the radio telescope. Daniel was confident what he had witnessed was a new player in the game, and a sophisticated one given the testing that took place. With no further activity for the moment, he decided it would be a good time to emerge from the confines of his bunker, allow his dog a chance to stretch and refresh, and return to his cabin to retrieve more items.

He and Dusty ascended the stairs and made their way over the embankment, passing one radio telescope, the observatory, then the other radio telescope and they descended down the pathway to the cabin. Once inside, Dusty made his way to his food and water bowl while Daniel began putting clothes in a duffel bag and collecting hygiene items from his bathroom.

The Grey Eagle approached the valley and, skimming the floor of the clouds, entered the operating area. On the unarmed drone, the Tactical Signals Intelligence Payload sensors captured a 360-degree aerial field of view, immediately identifying two moving targets in the area approaching the cabin.

Both Leonard and Ramin had brought their cell phone with them. Agent Burton called the quick reaction team headed to the area, still 30 minutes out, informing them of the two suspects closing in on the property. He then put the drone in a holding pattern, high above the mountain. The doctor did not have 30 minutes.

Agent Burton pulled out a tablet displaying the real-time aerial view feed from the Grey Eagle unmanned aircraft system, grabbed his tactical gear, and started down the mountain. The terrain was tough and there was no pathway. Agent Burton was not certain he could make it down in time to make a difference. Regardless, he began making his way down with full tactical gear.

Leonard was the first to arrive at Daniel's property. He spotted the road ahead and the fence line. He crossed the road and made his way along the fence to the north side of the property, eyeing the trees inside the property for the best cover to make his approach to the cabin. He did not notice the small red dots among the higher branches, triggering the motion sensors of the surveillance system. Alarms were going off in the bunker indicating an intruder, but no one was there to hear them.

He found a low area where a small wash descended the slope, and the fence crossed the wash. Leonard chose this as his crossing point. He swung the bolt cutters in front of him and contacted the wire fence, clipping one, then another, and then the last wire that crossed the dry wash. He hoped that Daniel had some animals that would be freed, as he fully intended to end their owner's life—a fair punishment for his transgressions against God.

The wash was a perfect entry point, leading directly into a thick outcropping of junipers. Creeping in, Leonard dropped the bolt cutters and pulled out his revolver. Slowly moving through the trees due to low visibility in the darkness, he made his way closer to the north side of the cabin. He could see the dimly illuminated exterior of the porch and entry stairs through the trees. He stopped, crouched, and searched for any activity in and around the structure. Hearing none, he continued to the edge of the tree line.

The lights were on in the cabin, so he could see clearly in through the large windows spanning both sides of the chimney. Leonard walked slowly to the stairs, gently shifting his weight onto each one in an attempt to avoid any creaking sound that might alert the cabin's inhabitants. He then approached the door and reached for the handle.

Daniel emerged from the bedroom down the hallway and had a direct line of sight through the window of the door. Seeing Leonard, Daniel instinctively froze in the doorway, duffle bag in his hand, and Leonard saw Daniel. Clutching his revolver, Leonard raised the gun and pointed it at Daniel through the glass windowpane of the door.

Nearby, Ramin made his way through the trees and clearings, tripping over uneven rocks barely visible in the darkness. Soon, he spotted the road ahead, and the fence line of a property on the opposite side. Cautiously, he crossed the dirt road and made his way along the fence to the south side of the property. He had not located the owner of the van, but he suspected that whoever it was meant to do harm to his target, and he could not let this happen until he got what he came for.

He found an ideal grouping of trees where the fence ran through the middle, then proceeded to twist the top two wires of the fence together using his tire iron. He climbed over the twisted wire, released the tension, and took the tire iron with him into Daniel's property. He crossed the drive headed uphill to the cabin and proceeded through the woods on the east side. As he approached the cabin, he saw movement from the north. A man emerged from the tree line and slowly crept up the stairs to the porch. *He's not the Dr. Sullivan in the video, so he must be the owner of the van,* he thought.

The man slowly made his way up the porch and around to the northside door of the cabin. Ramin was not going to let this man kill his mark. He pulled out his ghost gun, raised it toward the man, and walked briskly into the clearing in front of the cabin straight toward him. Rather than open the door as expected, the man raised his gun and pointing it inside the cabin. Without hesitation, Ramin fired, landing a round straight into the temple of the van owner.

His gun was not as lucky as the placement of its projectile. The burst of hot gasses was too much pressure for the type of printer filament used, shattering the barrel as the gun fired, sending shards of plastic in all directions, including one which sliced open a gash below his right eye.

Without even missing a stride, Ramin threw the remains of his gun on the ground and continued toward the man he had shot, who fell backwards and sideways onto the wooden porch. Ramin was tunnel focused now, thinking he must recover the man's revolver and finish his mission. The blood was beginning to soak in to

Leonard's ball cap by the time Ramin reached him and recovered the revolver on the porch.

Agent Burton made his way down the rough terrain on the side of the mountain, through brush and an uneven rock cropping, until he reached a drop-off that would surely cause more harm going down than around. Back tracking, he realized his best course was to follow the ridgeline down to the observatory in the distance. That would provide the best pathway to the cabin where his targets were headed.

He was mere yards away from the fence at the mountain top. Throwing his gear over top of the fence, he quickly cleared it and was standing in front of the observatory for the second time that day. A shot rang out, its echo bouncing off of the side of the mountain and reverberating into the valley. *I'm too late,* he thought, as he dropped his gear, grabbed his M-16, threw his tablet over his shoulder, and began tactically moving down the mountain toward the cabin.

Daniel, stunned into paralysis by the gunman at his door and equally shocked by the shot that rang out seemingly from nowhere dropped his duffle bag and retreated to his bedroom. *Today is not my day to die,* he thought, as he strategized how to keep that from happening. Someone out there was on his side, he surmised, by the abrupt change in circumstance.

Beneath his bed, a biometric safe was attached via cable to the bedframe. He pulled it out and opened it with his thumbprint, gaining access to credit cards, his passport, and—rummaging to the bottom—a 45-caliber 1911 handgun. Pulling out a loaded magazine, he loaded it into the gun and pulled the slide to insert one round in the chamber. He was not going to be a victim today.

There is no point in calling the police until this is over, he thought, keeping his phone in his pocket. Behind the blackout curtains was a set of windows that opened horizontally. Daniel considered making his exit, but feared he would be too exposed to whomever had just blown a hole in the side of the stranger on his porch. His best option was to remain in his bedroom with the door locked, while he sat on the floor on the opposite side of the bed with his 45-caliber pointed with both hands toward the door and window.

Ramin reached for the doorhandle that Leonard held just moments earlier, twisted it, and swung it open. Dusty, still in the kitchen at his food bowl, perked up a second time since the shot ring out. Seeing Ramin enter the cabin, he excitedly came over to meet their new guest, sniffing his legs and torso, tail wagging. Ramin backed out of the doorway, shooing the dog onto the porch where Leonard's body lay, then returned to the cabin and closed the door. Ramin continued through the great room, past the kitchen, and into the hallway toward the bedroom.

Agent Burton heard the eerie howls of the blue tick hound reverberate off of the side of the mountain, just as the gunshot had not long before. He made his way down the walkway to the back of the cabin and crouched next to the wall of the bedroom. He swung his tablet around and accessed the infrared camera imagery of the Grey Eagle drone.

At this point, he could see himself and another immobile figure on the opposing side of the same exterior wall. *This must be Dr. Sullivan,* he thought. There was a prone body, color fading on the opposite side of the cabin, and one more inside coming down the hallway toward the bedroom. *This must be the hostile.*

Agent Burton made his way around the south side of the cabin to the front and underneath the bedroom window, being cautious not to make himself known. He had been through the house earlier that day and knew that the doctor kept a personal weapon in his bedroom.

"Come out, Dr. Sullivan. I am not here to harm you," Ramin yelled. "I saved your life. That man was here to kill you. I just want you to give me a number."

Agent Burton had to think quickly before this person had an opportunity to cause harm again. He had already killed once just minutes ago. The agent needed a reference location inside the home that he could see on his aerial display. There wasn't enough time to get to the front door, if it was even unlocked. There were two windows in the hallway, with the bathroom opposite the outside in the middle between the two. This was halfway down the hall.

Kneeling, he placed the tablet in front of him and aimed for the center of the hallway. Ramin took a few more steps, stopping in front of the bathroom. Agent Burton placed three shots horizontally across the wall, mid-center of the windows.

Ramin felt a sting in his shoulder and a ringing in his ears at the same time. He involuntarily dropped the gun, unsure of what just happened to him. He had never been shot before, so it took a moment to come to his senses. He was in danger and his body felt the adrenalin rush, as he turned and ran back through the hallway and to the back door of the cabin. The shots had come through the front, so it was his best chance to escape. As he exited and began to run toward the trees to the south, Dusty barked and ran after him.

Realizing the threat was retreating, Agent Burton identified himself to Dr. Sullivan.

"Dr. Sullivan, I'm Agent Jose Burton from this morning. Are you okay?"

"Yes," Daniel yelled back, releasing his grip on the handle of the gun. He had not realized how tightly he was holding it.

"Stay where you are until we secure the suspect," Agent Burton said to Daniel.
"I'll let you know when it's okay to come out."

Agent Burton called the other agents and reported that there was one suspect down and another on his way through the woods heading eastward, as he watched him on the tablet stumbling to the white van. The agents put on their night vision goggles and spread out into the trees. Coordinating with Agent Burton, they closed in on the suspect, who was not moving very fast, apparently disoriented by the darkness and his injuries.

Ramin heard them coming but could not evade them. Attempting to run, he tripped, face planting into the rocks. They moved in quickly to apprehend Ramin and bind his hands with zip ties. Once they restrained him, they triaged his wounds, placed him in the back of the Humvee, and left for Flagstaff and the nearest hospital. A member of the team made a call to the

Coconino County Sherriff's office requesting both responders to the crime scene and an escort to the hospital.

Agent Burton called out to Dr. Sullivan again, "He's been captured. It is safe to come out now. Please secure your weapon before you do."

A wave of relief flooded over Daniel. He recognized the voice and was confident he was a genuine federal agent he could trust. He did as suggested, placing his firearm back into the safe, then returned it to its place under the bed. After a brief pause, soaking in the moment, he unlocked the bedroom door, opened it, and was face to face yet again with the second man to save his life that day.

Agent Burton escorted Daniel outside, careful to step around the blood splatter on the floor, then out the front door, purposely avoiding the northside entrance to the cabin. Most of his home was a crime scene now.

Dusty was waiting for him outside, excited to see his owner and did not seem to be bothered by the events that just transpired in front of him. Daniel was grateful he was not hurt in the ordeal and was there to comfort him through the aftermath.

The sheriff's deputies arrived around 30 minutes later. Agent Burton, along with another agent, Walters, who stayed behind to allow room in the Humvee, identified themselves and explained their version of events, which would be corroborated by surveillance video footage stored on a hard drive in the bunker. The video feed from the Grey Eagle was not going to be shared with county law enforcement unless absolutely necessary. Ramin's identity was still unknown, but it did not take long to locate and identify both vehicles and trace them to their respective owners.

The sheriff deputies took over the crime scene for now, took their statements, and released the three of them. Agent Burton guided the Grey Eagle back to the Camp Navajo Army National Guard Depot, where agents placed flares on the ground to light the path for a landing. Once the drone was safely on the ground, Walters made his way up to the communications site, packed up the gear, then drove Agent Burton's Humvee back to Daniel's property. After securing the bunker, Walters and Burton escorted Dr. Sullivan and Dusty to a hotel in Williams, Arizona, for the night. There would be a debrief with Dr. Sullivan in the morning. By then, lead Agent Devereux would be there, along with local agents to assume the investigation.

Chapter 20

The Regulation

After the testing at the U.S. Naval Observatory was completed, Levy discussed what comes next. A meeting was scheduled at the FCC the next morning to discuss the implications of the god frequency and what they could do to mitigate any negative repercussions. Antwan and Lauren were invited to attend, and they agreed.

The two agents accompanied the three of them back to a SUV, and they were on their way to a hotel for the night. Levy received a phone call and the two of them could only hear her side of the conversation. "Yes…I had suspected as much…No, that's not necessary, just continue to monitor for traffic and keep eyes on," she said, hanging up the phone. *She must be talking about Daniel,* thought Lauren.

Antwan, Lauren, and Agent Devereux were dropped off at the Washington Marriot Capitol Hill, just one block east of the FCC headquarters building. Levy handed Antwan and Lauren keys to a room with two queen beds on an upper floor with a city view. She had a rather liberal budget because of the significant amount of cryptocurrency she had recovered over the past few years.

Agreeing to meet at the lobby of the FCC headquarters promptly at 8:00 p.m. Eastern Time, Levy left them to their own devices—the first time since California the two of them no longer sensed a feeling of being in custody. Once they got to the room and relaxed for a moment, Lauren showered, and Antwan managed to get a reservation at 8:30 p.m. for the rooftop restaurant.

Yara, a Latin American inspired restaurant on top of the Washington Marriot, offered breathtaking views of Washington, D.C., and they arrived at the perfect time to see the sun set across the city. At the side of the building were the tracks leading to and from Grand Central Station. The sound of trains periodically emanated upward to the terrace from the ground below was somehow a positive addition to the restaurant's ambiance.

It feels like a strange but proper time to celebrate, thought Antwan, given the vindication of their discovery, combined with the seizure of his transceivers and the forced suppression of their speech. They decided to indulge in a short-rib farrotto paired with a bottle of Carmenere.

The pink and orange hue in the sky faded to purple and gray and gave way to the night sky of the city before Antwan and Lauren found the bottom of the bottle. They made their way down to their room, thankful there was no reason to leave the building that night. The day had been a whirlwind from before dawn to after dusk.

Back at their room, Antwan opened the curtains of the large windows overlooking the city's skyline.

Lauren laid down on the bed next to him, nestled her head on to Antwan's shoulder, and promptly fell asleep. Trapped, Antwan had time to contemplate the events of the day and mentally prepare for the meeting tomorrow.

The morning sunlight rudely woke them up before 6:00 a.m. local time. The room got brighter, but it wasn't until the sun crested the horizon, sending beams over the rooftops of neighboring buildings and into the uncovered windows of their hotel room, that Antwan and Lauren began to stir.

They eventually got out of bed and dressed for another day in their guided tour of classified government operations. Checking out of the room, the two donned their backpacks and began walking the streets of the NoMa district. Across the street from the FCC headquarters was the National Public Radio (NPR) headquarters. Home to the desk of Bob Boilen, the highly sought-after venue that hosts "Tiny Desk Concerts" on NPR.

Approaching 8:00 a.m., the two made their way through the revolving doors and entered the main entrance of the FCC headquarters. One of Levy's associate agents waiting for them in the lobby explained that Agent Devereux could not attend today's meeting.

Lauren found it odd that Agent Devereux was absent that morning, surmising something more important had taken place that required her attention. Lead agents don't typically work multiple unrelated cases, so whatever was going on that required her absence must have to do with the frequency as well.

He escorted them into an elevator to the 11th floor, down a hall, and to a corner conference room with floor-to-ceiling windows on two walls overlooking the NPR headquarters across the street adorned with satellite dishes along the edge of its lower roof, obviously placed there for their aesthetics as much as for their function. A tower crane loomed in the distance, giving an indication of the constant evolution of real estate in the area.

The conference room was obviously one of the more important spaces at the headquarters, as indicated by its prime location and large, leather chairs around an oval conference table. Screens along the short interior wall had screens for video conferencing and presentations.

This meeting was intended to cast a wide net, with participants from NASA, DARPA, DOD, DOE, JPL, FERC, FCC, and several other government agencies and departments, including intelligence services and selected unnamed government partners, though why these last two participants attended was not clear.

Leading the meeting was the chief of staff and legal advisor for one of the FCC commissioners. The meeting began with a presentation of the testing that had been performed by the U.S. Naval Observatory the day before. Indications were corroborated by a second receiver somewhere in the Western U.S.—a complete shock to Antwan and Lauren. That certainly wasn't a part of their testing and must have been determined later. They had both assumed that Dr. Sullivan's radio telescopes had been disabled or seized as well, though it would not have been practical to just pick up and leave with two large dish radio telescopes.

The next phase of the meeting was a discussion on the potential uses and benefits to the god frequency. First and foremost were potentially instantaneous interstellar communications that opened real-time communications for space travel and exploration.

Next, someone from the Department of Energy or the Federal Energy Regulatory Commission proposed the potential for a loss-free method to transmit power over an infinite distance. This prospect also appealed to the space agencies. Significant research was going to be done to explore these options. Any further uses, and it was apparent each party at the table had them, were held close to their chests.

The discussion moved on to a discussion of concerns about the use of the frequency and security risks. Levy's rationale for confiscating the transceivers and limiting access to the actual frequency number was the first item of discussion. Misuse of the frequency could destroy any receiver, and no protective device could be implemented as a time curve does not exist. As such, it was imperative to keep the frequency value and technology to access the frequency out of the wrong hands.

Other uses discussed were triggering weapons or accessing areas inside shielded compounds, fortified bunkers, or even Faraday cages.

Locating an unauthorized broadcast was another dilemma. It would be difficult to trace a signal without a reference to its origin in the signal.

This was the one time Antwan addressed the group, "A frequency source could potentially be traced by its phase noise. It is highly unlikely that someone can transmit only on this precise frequency without the broadcast spilling over to adjacent frequencies, which transmit like any other frequency. If there are sensitive enough receivers within the broadcast range looking for transmission on the side frequencies, the location of a transmission on the god frequency can be found."

Many in the room seemed impressed by his input. A young man with dreads, dressed more casually than most in the room and seemingly out of place, was providing valuable input into the conversation. They were already aware of who he was and what he had built and discovered but they did not respect his participation in the conversation until then.

Finally, the topic turned to regulation. The FCC already regulates the frequency range where the god frequency resides. More specific regulation was needed to prohibit transmissions on the precise frequency itself. Any normal amateur radio would not be able to transmit on the frequency because it is not precise enough for them.

The dilemma posed with regulating it had to be defined. To do this, they would have to give the frequency, or a frequency range, to the public, which meant giving it to foreign adversaries or bad actors who would not be deterred by an FCC ban in the first place.

The next best option was controlling the technology to access the frequency. A precedent was already established with export controls on equipment that can be used for military applications or nuclear technologies, both power and weaponry. The FCC would go forward and propose strict export controls on technologies that could be used to make precision transmitters accurate to one septillionth of a Hertz.

After the conference was over, Levy's agent handed Lauren and Antwan commercial airline tickets back to San Diego. Their foray into government regulatory space was over as quickly as their foray into government military operations. It had been an eye-opening experience, but now they had to go back to their lives and their parents to explain why their houses had been raided by government agents.

Chapter 21

The Tech Sector

"Okay, thank you, Agent Devereux," said the Director of National Intelligence, ending the phone call. Levy had just acquired the frequency file Antwan had sent to Daniel, a direct request from the Director. In fact, the case and Levy's selection for it were requested directly by her. There was much development work to be done and a short time to do it. Before regulations and international agreements were made, domestic technology companies needed to know the specifications required to access the frequency and get to work creating and improving the technology for real-world applications.

Raytheon and other key partnerships through DARPA developed transceivers of in classified contracts for Department of Defense applications. NASA explored its technologies for use in satellites and on the International Space Station for research and space exploration. Research universities and laboratories in partnership with the government developed projects for discovery and real-world applications.

Unfortunately, amateur radio applications were shut out of the initial conversation. The level of precision would initially be cost prohibitive to most amateur radio operators anyway, but the impending access restrictions on the frequency would have to exclude the public from buying off-the-shelf equipment that could broadcast on it. Besides, all the major tech companies producing amateur radio equipment were not domestically owned. The unintended consequences of this discovery were still unknown and needed a cautious approach.

Withholding the frequency would create a sore spot with the international scientific community. This was potential game changer to the understanding of the universe, acknowledged the director, referencing Dr Sullivan's speech. She would know. She had a degree in physics. But that didn't change the sensitive nature of the discovery.

The frequency would inevitably get out, but the United States needed to get ahead of the game in research and technological development. If they were to allow the frequency to get out immediately, adversaries would have an equal opportunity to find ways to use it against the U.S., and it was the director's responsibility to be one step ahead.

Fortunately for the director, even though she was a career civil servant, she had connections in the tech sector. Her first contact was Master-Stone, a software company delivering Artificial Intelligence (AI) products using high-power semiconductor chips originally meant for video graphics cards then repurposed for AI. Master-Stone adopted its name from J.R.R. Tolkien's *Lord of the Rings* trilogy, the Master-Stone being the most powerful of the palantíri or seeing stones.

The director tasked them with creating micro transmitters accurate to one yoctohertz that could transmit data on an extremely low wattage signal, undetectable to the standard monitoring device, and a sensitive receiver that could pick it up. This was an ambitious task, but it could be a game changer in the intelligence world. Imagine having an unlimited range of data transmission that was undetectable with no dead zones.

Antwan, understanding the sensitive nature of the discovery, nevertheless retained a lawyer to seek the return of his transceivers. The government did not have a lawful reason to confiscate them, and a judge ruled in his favor, issuing a court order for the return of the transceivers, design documentation, all amateur radios seized from Antwan and Lauren, and all computers seized from Antwan, Lauren, and Dr. Sullivan.

Yaesu Musen Co., Ltd., a popular amateur radio manufacturer, made an offer to Antwan for his transceiver design. Coordinating within the confines of the new regulations, Yaesu USA, agreed to outsource manufacturing of precision transceivers for the United States market to a company named Henry Radio, Inc., thus adhering to the strict export and manufacturing controls of the new agreement. In addition, receive function would be maintained, but transmit function would be blocked on the god frequency.

The offer made by Yaesu USA was a generous one, anticipating multiple offers from other amateur radio manufacturers. Antwan did not need to wait for other offers to accept.

Chapter 22

The Academicians

As Lauren, Daniel, and Antwan had hoped, news of the god frequency traveled fast through academia. It did not take long for theories to be made, as technological research advancements focused on precision of frequency to a target of a septillionth of a Hertz, followed by government-university research projects. Technology companies had been involved in development since before government regulation.

One theoretical physicist proposed that rather than violate the commonly held belief that nothing can go faster than the speed of light, theory of relativity aside, that perhaps the god frequency "tuned in" to dark energy like some sort of resonant frequency. Imagine a tightly packed structure, like a carbon nucleus in a diamond. If you began to vibrate that single nucleus, it would, in turn, move the adjacent nuclei proportionally to the vibration and the diamond would vibrate. Dark energy could be acting like the diamond.

NASA created research projects coordinating with Master-Stone, who had already developed a micro-transceiver accurate to a yoctohertz, to be placed into satellites, and radio manufacturers to incorporate a precision-frequency transceivers into spacecraft and the International Space Station.

SETI worked with the tech industry to upgrade receivers of its radio telescopes, in much the same fashion as Dr Sullivan had, and they began a constant surveillance, coordinating with known entities and monitoring for phase noise to locate the origin of detected signals.

Chapter 23

The Recognition

After a few weeks, the frenzy of phone calls, emails, and text messages dwindled to the occasional latecomer to the news of the god frequency. Antwan and Lauren did their best to screen them and avoid the discussion. The existence of the god frequency had become common knowledge in the academic world, and those who discovered it, but a bit of a legend or myth to the general public because it was intangible, incongruent with the rest of the observational world, and not understood. The gag order they had received no longer applied, save for the specific frequency and design details of transceiver precision. It became a sensitive topic to them not only because Antwan's revolutionary technology was confiscated but also because of the ordeal that Dr. Sullivan went through while they were on an all-expenses paid field trip to the Capital.

Their other amateur radios were returned, and Antwan and Lauren's interactions more or less returned to their normal, meeting to talk on their frequency. Antwan packed to return to the USC's main campus to begin his Master's program, while Lauren's PhD program continued. From that distance, their normal frequency would be out of range. They had grown closer, even if their life goals kept them separated, and they decided to keep in touch with traditional methods, like email, texts, and phone calls since they no longer had a chest freezer with unlimited range.

Dr. Sullivan kept them apprised of some of the activity he was seeing on the frequency. His computers were returned following formal regulation approvals and ratified treaties. The United States government still kept eyes on him and his property, but like listening to any other amateur radio frequency, there was no harm in receiving the signal. Besides, he still had possibly the best spyware protection that money could buy. Lauren and Antwan made a deal to continue their studies through the following summer, with a calculation that both could graduate at the same time at the end of the next fall semester.

Antwan returned to Oceanside one day the fall before graduation. Lauren came over to spend the weekend, as she had every time they could get together the last six months.

Early Sunday morning, before the sky began to change color, Antwan's phone began to buzz. The barrage of phone calls had ceased months before, and even the trickle had faded into obscurity, as the news cycle caught the sails of the newest crazed story. He was up this time, and saw a rarity on his phone. It was a foreign combination of numbers he did not recognize: 46 08 663 09 20.

Out of curiosity, he did something he seldom did and slid the green button of his phone on an unknown number.

"Hello," said Antwan.

"Hello," said a male voice on the other end of the phone, speaking English with a Scandinavian accent.

"Is this Antwan Richard?" asked the caller.

"Yes, can I help you?" he replied.

"This is Dr. Sigurd Westermark of the Royal Swedish Academy of Sciences," the called replied. "Antwan Richard, you, Lauren McCartney, and Dr. Daniel Sullivan have been selected as the laureates of the Nobel Peace Prize in Physics for the current year for your work leading to the discovery and presentation of the god frequency."

Antwan was in shock. This sounded like a joke, if not for the early hour and the phone number on the caller ID. It was the first time he had heard reference to god frequency since the previous year.

The caller continued, "The Nobel Prize Award ceremony will take place on December 10th in Stockholm. Additional details will be communicated as they become available. Congratulations, Antwan! We look forward to seeing you then."

"Thank you very much," he said, still processing the information and the implications.

Antwan shook Lauren awake, "Lauren, Lauren!"

Still not fully awake, she looked at him incredulously.

"You will not believe what just happened," he told her.

Her incredulous look did not change. She at this point was not amused. *He needs the equivalent of another discovery defying the known laws of physics to impress me at this point,* she thought. And she knew them fairly well, considering she was on the verge of her PhD in geophysics.

"We, along with Dr. Sullivan, have been awarded the Nobel Prize in Physics!" he exclaimed.

The squint-eyed expression on her face remained unchanged. He was beginning to think it was permanent. Clearly, she was still processing.

"For our work on discovering the god frequency," he continued. "He said that!"

She sat up, starting to believe his words and transitioned from processing to shock.

"If you don't believe me, surely they will be calling you, too. And Dr. Sullivan."

Lauren's phone began to buzz, but it was not a foreign number. It was Dr. Sullivan.

"Did you get the phone call," he asked.

"Are you talking about the Nobel?" she asked.

"Yes," he cried out in amazement.

"No. But Antwan did," she continued. "Remember, I changed my number a couple of months ago. Perhaps they didn't get the update."

She could not think of who might possibly have nominated them, or how the others' contact information was found. Statutes restrict disclosure of information on nominations and nominators to the Nobel Foundation for 50 years, so there would be no formal way to figure that one out.

Lauren called her parents to see if they had heard anything. They had not, but they were overjoyed with the news. Typically, the laureates find out just moments before it is publicly announced. She and Antwan kept their eyes on the news for the formal announcement. At breakfast, Lauren had the honor to tell Byron and Marie that they were parents of a Nobel laureate. Before they had finished their plates, the call came in to Lauren as well.

To Lauren, it was a surreal moment, even given nearly an hour to prepare herself.

Incredible, she thought, that the two of them, not yet even doctors, had achieved the pinnacle award in Physics. *But, then again, we identified and shared perhaps the most consequential discovery in physics this millennia,* she boasted to herself. Most definitely in the last year.

In early December, classes were finally over for Antwan and Lauren for the fall semester. Both had arranged to take their finals early so they could travel to Stockholm for the Nobel Prize award ceremony. Lauren's father said it was the "noble thing to do." She had already defended her dissertation and, thus, technically a was a doctor. Antwan had yet to defend his thesis, but the department head graciously deferred the task until after the trip.

It was a family affair. Lauren's parents, Antwan's parents, Antwan's brother, and the two of them boarded a plane to London's Heathrow Airport where they had a layover, then a connecting flight to Stockholm, Sweden. A total travel time of 17 hours and 30 minutes. Luckily, they had a few days to recover before the award ceremony.

Traveling together in Europe was an experience. For all the differences of the two families that could be causes for division in the United States, were counteracted by their cultural similarities when overseas. As diverse as they were, they were seen collectively as Americans, and it elicited a sense of inward pride to Antwan. *This must be what my father felt when he went abroad in the military,* he thought. They represented their country, whether they intended to or not.

They were like fish out of water in the Scandinavian winter, though. Daily high temperatures were just above freezing, and everyday lows below that, adding to the foreign atmosphere of the experience. They purposely traveled without winter clothes, choosing instead to purchase locally, so at least their clothes didn't aesthetically identify them as tourists to the locals.

When the day came and they arrived at the Stockholm Concert Hall, Antwan and Lauren departed from their families and met up with Dr. Sullivan. The Nobel laureates were ushered into a separate area on stage in the center of the concert hall for the ceremony. Swedish royalty entered the room and assumed their positions in front of their chairs on the opposite side of the stage in front of members of the Nobel Foundation.

To Lauren, the atmosphere of the Concert Hall was cathartic, the vibrations of the stringed instruments transiting across the room and back again. If there was an occasion to dress in strict formal attire, this was it. The men were uniformly dressed in white tie and tails, while all women wore evening gowns.

Antwan, soaking it all in, imagined the sound wave propagation through the room, attempting to understand the acoustic considerations in the design of the space.

Dr. Sullivan, was more nostalgic, reflecting on his career and the choices of his life that included a chance encounter with Antwan and Lauren that led him to sitting on this stage on the anniversary of Alfred Nobel's death.

The presentation speech for the Physics Prize was given by Professor Sigurd Westermark, Member of the Royal Swedish Academy of Sciences; Chair of the Nobel Committee for Physics:

"Your Majesties, Your Royal Highnesses, Esteemed Nobel Prize Laureates, Ladies and Gentlemen,

'Somewhere, something incredible is waiting to be known.'

These are words by Dr. Daniel Sullivan, and they illustrate the fact that there are incredible things in the universe, just underneath the surface waiting to be discovered by those who look.

In studying physics, we strive every day to understand the observable world through defining the natural laws of the universe.

In Albert Einstein's early work on the photoelectric effect, he proposed that light behaves as particles called photons, with each photon carrying a discrete amount of energy directly related to its frequency, giving an explanation for aspects of the photoelectric effect that could not be explained by the wave theory of light alone for which Albert Einstein received the Nobel Prize in Physics in 1921.

This understanding of light waves laid the groundwork for quantum mechanics. However, even Einstein believed the universe was static until 1931. Some attribute his change in perspective to American astronomer Edwin Hubble, showing him observations of redshift in the light emitted by far away nebulae today known as galaxies.

Edwin Hubble's work was expanded upon by another great discovery, when in the late 1990s, Saul Perlmutter, Brian Schmidt, and Adam Riess, while observing Type 1A supernovae, discovered the objects appeared to move faster. Their observations led to a conclusion that not only is the universe expanding, but that its expansion is speeding up. For this work, these three individuals were awarded the Nobel Prize in Physics in 2011.

Douglas Hemme

In order to explain the phenomenon of an expanding universe, a theory of dark energy was formulated. It is estimated that dark energy, an unknown force that is believed to be responsible for the universe's accelerating expansion, makes up nearly 70% of the entire universe's mass and energy, overcoming gravity to pull the universe apart.

This year's Nobel Prize in Physics is focused on small frequency scales, and to be precise, on yoctohertz. One septillionth of a cycle per second. This was the level of precision required to discover an energy wave that defies our current knowledge of the universal laws of physics.

Colloquially referred to as the 'god frequency,' an energy wave traveling on this frequency is everywhere all at once. Imagine flipping a light switch. The room fills with light at a rate imperceivable to the human eye. Imagine, then, that the light switch is a broadcast on the god frequency. On flipping the light switch, the entire universe lights up, light years of distance, at a rate of speed imperceivable to the most precise measurements of time in existence today.

This discovery is taking us closer to understanding the interaction of matter, energy, dark energy, and dark matter. It is the next step in understanding the universe, the next step in understanding quantum mechanics, and the next step for modern-day physics.

Emeritus Professor Sullivan, Dr. McCartney, and Mr. Richard, you have been awarded the 2026 Nobel Prize in Physics for the development of technology leading to the discovery, demonstration, and exhibition of the god frequency. It is an honor and a privilege to convey to you, on behalf of the Royal Swedish Academy of Sciences, our warmest congratulations.

I now ask you to step forward to receive the Nobel Prize from the hands of His Majesty the King."

It was a strange introduction for Antwan and Lauren into careers in science. Like medal-earning Olympians, this Nobel Prize would follow and define them for years to come for better or for worse.

For Dr. Sullivan, it is a fitting culmination of a decades-long career dedicated to teaching generations of young students the modern-day understanding of physics, along with the history and evolution of its understanding.